Computational Biomechanics for Medicine

Grand R. Joldes • Barry Doyle • Adam Wittek
Poul M.F. Nielsen • Karol Miller
Editors

Computational Biomechanics for Medicine

Imaging, Modeling and Computing

Springer

Editors
Grand R. Joldes
Intelligent Systems for Medicine Laboratory
School of Mechanical and Chemical Engineering
The University of Western Australia
Perth, WA, Australia

Adam Wittek
Intelligent Systems for Medicine Laboratory
School of Mechanical and Chemical Engineering
The University of Western Australia
Perth, WA, Australia

Karol Miller
Intelligent Systems for Medicine Laboratory
School of Mechanical and Chemical Engineering
The University of Western Australia
Perth, WA, Australia

Barry Doyle
Intelligent Systems for Medicine Laboratory
School of Mechanical and Chemical Engineering
The University of Western Australia
Perth, WA, Australia

Poul M.F. Nielsen
Auckland Bioengineering Institute
The University of Auckland
Auckland, New Zealand

ISBN 978-3-319-28327-2 ISBN 978-3-319-28329-6 (eBook)
DOI 10.1007/978-3-319-28329-6

Library of Congress Control Number: 2016938222

Printed on acid-free paper

This Springer imprint is published by Springer Nature
The registered company is Springer International Publishing AG Switzerland

Preface

The field of computational biomechanics continues to progress toward clinically relevant simulations. Personalized medicine will play a major role in the future of healthcare, and personalized computational simulations have a clear part to play. We, the computational biomechanics community, are making real advancements toward this ultimate goal of comprehensive patient-specific modeling, but there is of course much more work to do yet.

The first volume in the *Computational Biomechanics for Medicine* book series has been published in 2009. Since then, the book has become an annual reference for the community to read about the latest advancements in the field. The book series provides an opportunity for specialists in computational sciences to describe their latest results and discuss the possibility of applying their techniques to computer-integrated medicine.

This seventh volume in the *Computational Biomechanics for Medicine* book series comprises 18 of the latest developments in solid biomechanics, vascular biomechanics, and brain biomechanics, from researchers in Australia, New Zealand, USA, UK, Switzerland, Scotland, France, and Russia. Some of the exciting topics discussed are:

- Tailored computational models
- Traumatic brain injury
- Soft-tissue mechanics
- Medical image analysis
- Clinically relevant simulations

The *Computational Biomechanics for Medicine* book series does not only provide the community with a snapshot of the latest state of the art, but more importantly, when computational biomechanics and patient-specific modeling is a mainstay of personalized healthcare, it will serve as a key reminder of how the field has developed over the years.

We would like to thank the authors for submitting high-quality work and the reviewers for helping with paper selection.

We hope you enjoy this year's edition.

Perth, WA, Australia — Grand R. Joldes
Perth, WA, Australia — Barry Doyle
Perth, WA, Australia — Adam Wittek
Auckland, New Zealand — Poul M.F. Nielsen
Perth, WA, Australia — Karol Miller

Contents

Contributors

Christopher J. Arthurs Division of Imaging Sciences and Biomedical Engineering, King's College London, King's Health Partners, St. Thomas' Hospital, London, UK

Chloé Audigier Inria Sophia-Antipolis, Asclepios Research Group, Sophia-Antipolis, France

Siemens Corporate Technology, Imaging and Computer Vision, Princeton, NJ, USA

T.I. Avsievich Tambov State Technical University, Tambov, Russia

Nicholas Ayache Inria Sophia-Antipolis, Asclepios Research Group, Sophia-Antipolis, France

Habib Y. Baluwala Auckland Bioengineering Institute, University of Auckland, Auckland, New Zealand

Maxime Berg Department of Fluid Mechanics and Hydraulics, ENSEEIHT, Toulouse, France

Philippe Büchler Institute for Surgical Technology and Biomechanics, University of Bern, Bern, Switzerland

Dorin Comaniciu Siemens Corporate Technology, Imaging and Computer Vision, Princeton, NJ, USA

Noel Conlisk Centre for Cardiovascular Sciences, University of Edinburgh, Edinburgh, UK

Michael T. Cooling Auckland Bioengineering Institute, University of Auckland, Auckland, New Zealand

Clément Coutey Department of Fluid Mechanics and Hydraulics, ENSEEIHT, Toulouse, France

Hervé Delingette Inria Sophia-Antipolis, Asclepios Research Group, Sophia-Antipolis, France

Michele Diana IHU, Strasbourg, France

Barry Doyle Vascular Engineering, School of Mechanical and Chemical Engineering, The University of Western Australia, Perth, WA, Australia

T.M. Ecker Department of Orthopaedic Surgery, Inselspital, University of Bern, Bern, Switzerland

Nele Famaey Biomechanics Section, KU Leuven, Heverlee, Belgium

C. Alberto Figueroa Department of Surgery, University of Michigan, Ann Arbor, MI, USA

Department of Biomedical Engineering, University of Michigan, Ann Arbor, MI, USA

Division of Imaging Sciences and Biomedical Engineering, King's College London, King's Health Partners, St. Thomas' Hospital, London, UK

S.V. Frolov Tambov State Technical University, Tambov, Russia

F. Germain LBM/Institut de Biomécanique Humaine Georges Charpak, Arts et Metiers ParisTech, Paris, France

Seongmin Ha Weill Cornell Medical College, New York, NY, USA

Harvey Ho Auckland Bioengineering Institute, University of Auckland, Auckland, New Zealand

Lyam Hollis Centre for Cardiovascular Sciences, University of Edinburgh, Edinburgh, UK

Peter R. Hoskins Centre for Cardiovascular Sciences, University of Edinburgh, Edinburgh, UK

Peter Hunter Auckland Bioengineering Institute, University of Auckland, Auckland, New Zealand

Grand R. Joldes Intelligent Systems for Medicine Laboratory, The University of Western Australia, Perth, Australia

Ali Kamen Siemens Corporate Technology, Imaging and Computer Vision, Princeton, NJ, USA

Lachlan Kelsey Vascular Engineering, School of Mechanical and Chemical Engineering, The University of Western Australia, Perth, WA, Australia

Bishesh Khanal Inria Sophia Antipolis Méditerranée, Asclepios Research Project, Sophia Antipolis, France

Michael Kistler Institute for Surgical Technology and Biomechanics, University of Bern, Bern, Switzerland

Andreas Kristen Vascular Engineering, School of Mechanical and Chemical Engineering, The University of Western Australia, Perth, WA, Australia

Alesio L. Lanzara Intelligent Systems for Medicine Laboratory, The University of Western Australia, Perth, Australia

Mao Li Intelligent Systems for Medicine Laboratory, The University of Western Australia, Perth, Australia

Yi Chung Lim Auckland Bioengineering Institute, University of Auckland, Auckland, New Zealand

L. Liu Institute for Surgical Technology and Biomechanics, University of Bern, Bern, Switzerland

David S. Long Auckland Bioengineering Institute, University of Auckland, Auckland, New Zealand

Marco Lorenzi University College London, Translational Imaging Group, London, UK

INRIA Sophia Antipolis Méditerranée, Asclepios Research Project, Sophia Antipolis, France

Duane T.K. Malcolm Auckland Bioengineering Institute, University of Auckland, Auckland, New Zealand

Tommaso Mansi Siemens Corporate Technology, Imaging and Computer Vision, Princeton, NJ, USA

Sue R. McGlashan Auckland Bioengineering Institute, University of Auckland, Auckland, New Zealand

Viorel Mihalef Siemens Corporate Technology, Imaging and Computer Vision, Princeton, NJ, USA

Karol Miller Intelligent Systems for Medicine Laboratory, The University of Western Australia, Perth, Australia

Institute of Mechanics and Advanced Materials, Cardiff School of Engineering, Cardiff University, Cardiff, Wales, UK

James K. Min Weill Cornell Medical College, New York, NY, USA

Martyn P. Nash Auckland Bioengineering Institute, University of Auckland, Auckland, New Zealand

Department of Engineering Science, University of Auckland, Auckland, New Zealand

Poul M.F. Nielsen Auckland Bioengineering Institute, University of Auckland, Auckland, New Zealand

Department of Engineering Science, University of Auckland, Auckland, New Zealand

Pankaj Pankaj School of Engineering, University of Edinburgh, Edinburgh, UK

Tiziano Passerini Siemens Corporate Technology, Imaging and Computer Vision, Princeton, NJ, USA

Torben Pätz Fraunhofer MEVIS, Bremen, Germany

Xavier Pennec INRIA Sophia Antipolis Méditerranée, Asclepios Research Project, Sophia Antipolis, France

David M. Pierce Department of Mechanical Engineering, University of Connecticut, Storrs, CT, USA

H. Pillet LBM/Institut de Biomécanique Humaine Georges Charpak, Arts et Metiers ParisTech, Paris, France

Raoul Pop IHU, Strasbourg, France

A.Y. Potlov Tambov State Technical University, Tambov, Russia

Tobias Preusser Fraunhofer MEVIS, Bremen, Germany

Jacobs University Bremen, Bremen, Germany

S.G. Proskurin Tambov State Technical University, Tambov, Russia

Saikiran Rapaka Siemens Corporate Technology, Imaging and Computer Vision, Princeton, NJ, USA

Mauricio Reyes Institute for Surgical Technology and Biomechanics, University of Bern, Bern, Switzerland

Neil Roberts Clinical Research Imaging Centre, University of Edinburgh, Edinburgh, UK

G. Rochcongar LBM/Institut de Biomécanique Humaine Georges Charpak, Arts et Metiers ParisTech, Paris, France

P.Y. Rohan LBM/Institut de Biomécanique Humaine Georges Charpak, Arts et Metiers ParisTech, Paris, France

P. Rouch LBM/Institut de Biomécanique Humaine Georges Charpak, Arts et Metiers ParisTech, Paris, France

S. Schumann Institute for Surgical Technology and Biomechanics, University of Bern, Bern, Switzerland

Lars Ole Schwen Fraunhofer MEVIS, Bremen, Germany

K.A. Siebenrock Department of Orthopaedic Surgery, Inselspital, University of Bern, Bern, Switzerland

Matthew Sinclair Intelligent Systems for Medicine Laboratory, The University of Western Australia, Perth, Australia

W. Skalli LBM/Institut de Biomécanique Humaine Georges Charpak, Arts et Metiers ParisTech, Paris, France

Jos Vander Sloten Biomechanics Section, KU Leuven, Heverlee, Belgium

Luc Soler IHU, Strasbourg, France

IRCAD, Strasbourg, France

Vukašin Štrbac Biomechanics Section, KU Leuven, Heverlee, Belgium

Peng Sun Weill Cornell Medical College, New York, NY, USA

Elham Taghizadeh Institute for Surgical Technology and Biomechanics, University of Bern, Bern, Switzerland

Lauren Thomas-Seale Centre for Cardiovascular Sciences, University of Edinburgh, Edinburgh, UK

P. Thoreux LBM/Institut de Biomécanique Humaine Georges Charpak, Arts et Metiers ParisTech, Paris, France

Erich Wintermantel Institute of Medical and Polymer Engineering, Faculty of Mechanical Engineering, University of Technology Munich, Garching, Germany

Adam Wittek Intelligent Systems for Medicine Laboratory, The University of Western Australia, Perth, Australia

Guanglei Xiong Weill Cornell Medical College, New York, NY, USA

G. Zheng Institute for Surgical Technology and Biomechanics, University of Bern, Bern, Switzerland

Haoyin Zhou Weill Cornell Medical College, New York, NY, USA

Part I
Biomechanics of Solids

Chapter 1
Computer Assisted Planning of Periacetabular Osteotomy with Biomechanical Optimization: Constant Thickness Cartilage Models vs. Patient-Specific Cartilage Models

L. Liu, T.M. Ecker, S. Schumann, K.A. Siebenrock, and G. Zheng

1.1 Introduction

Periacetabular osteotomy (PAO) is an effective approach for surgical treatment of hip dysplasia in young adults [1]. The aim of PAO surgery is to increase acetabular coverage of the femoral head and to reduce contact pressures by realigning the hip joint. It was reported [2] that PAO planning approach is mainly based on two types of optimization strategies which are morphology-based and biomechanics-based optimization, respectively.

Both 2D and 3D images have been used in morphology-based planning of PAO surgeries. Clohisy et al. [3] reported that hip specialists identify important radiographic features of the hip on plain radiographs for diagnosis of pathological hips. As it is known that one dimensional information is missing with X-ray imaging due to the nature of projection, CT has been used as a more accurate and informative way of analyzing the morphology of the hip. Klaue et al. [4] proposed a CT evaluation method estimating coverage and congruency of hip joint. They made a topographical map of the acetabulum and the femoral head from the cross section images of CT scan and calculated the acetabular coverage. More recently, Dandachli et al. [5] described a new CT-based evaluation method for dysplastic hip from the weight bearing surface point of view. All these previously introduced methods [4, 5] quantify the morphology of dysplastic hips with pure geometrical measurements.

L. Liu (✉) • S. Schumann • G. Zheng
Institute for Surgical Technology and Biomechanics, University of Bern, Stauffacherstrasse 78, Bern 3014, Switzerland
e-mail: li.liu@istb.unibe.ch; guoyan.zheng@istb.unibe.ch

T.M. Ecker • K.A. Siebenrock
Department of Orthopaedic Surgery, Inselspital, University of Bern, Bern 3010, Switzerland

G. Joldes et al. (eds.), *Computational Biomechanics for Medicine*,
DOI 10.1007/978-3-319-28329-6_1

The other type of planning strategy is based on biomechanics optimization. Zhao et al. [6] conducted a 3D finite element (FE) analysis of acetabular dysplasia. The effects of dysplasia and PAO were both investigated by analyzing the change of Von Mises stress in the cortical bone before and after surgery. They showed that the PAO may be beneficial. One limitation of this method lies in the fact that the acetabular dysplastic model representing different levels of severity of dysplasia were generated by deforming the acetabular rim of a normal hip. Thus, it ignores the influence of the abnormal acetabulum of the real dysplastic hip. In contrast, the computer assisted Biomechanical Guidance System (BGS) introduced by Armand et al. [2] combines geometric and biomechanical feedback with intra-operative tracking to guide the surgeon through the PAO procedure. During the planning stage, the PAO planning computes contact pressures via Discrete Element Analysis (DEA) in order to suggest a reorientation of the acetabulum that minimizes simultaneous peak contact pressure in sitting, standing, and walking positions [7]. Recently, Zou et al. [8] developed a 3D FE simulation of PAO and validated their method on five models generated from CT scans of dysplastic hips. The acetabulum of each model was rotated in 5° increments in the coronal plane from original lateral center edge (LCE) angle, and the relationship between contact area and pressure and Von Mises stress in the femoral and pelvic cartilage were investigated until the optimal position for the acetabulum following PAO was found. However their virtual PAO procedure was performed with commercial FE analysis software Abaqus (Dassault Systèmes Simulia Corp., USA) which does not have a precise virtual reorientation planning tool for an accurate quantification of patient-specific 3D hip joint morphology.

Previously, we have developed a computer assisted planning and navigation system for PAO [9], which allows for not only quantifying the 3D hip morphology with geometric parameters such as acetabular orientation (expressed as inclination and anteversion angles with respect to the so-called Anterior Pelvic Plane (APP) [10]), LCE angle and femoral head coverage for a computer assisted diagnosis of hip dysplasia but also virtual PAO surgical planning and simulation (Fig. 1.1). In this paper, based on this previously developed PAO planning system, we developed a patient-specific 3D FE model to estimate the optimal acetabulum reorientation for planning PAO. One key factor that may influence the biomechanical optimization results is related to the cartilage models used in the FE simulation. In the literature, both constant thickness cartilage models [8] and patient-specific cartilage models [11] have been suggested. However, little is known about how different cartilage models used in the FE simulation may further affect the biomechanical optimization-based PAO planning. Our aim is to investigate the influence of these two different cartilage models on the biomechanical optimization results.

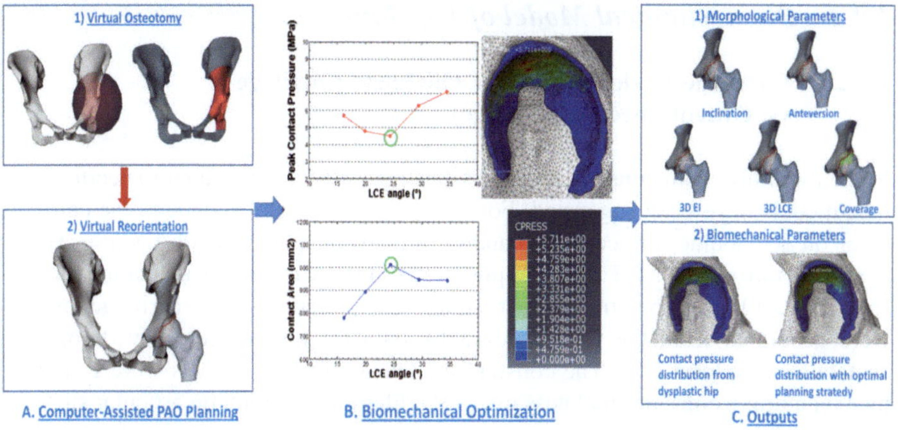

Fig. 1.1 Schematic view of our computer assisted planning of PAO with biomechanical optimization. (**a**) computer assisted morphology-based PAO planning. Virtual osteotomy operation is done with a sphere, whose radius and position can be interactively adjusted, and virtual reorientation operation is done by interactively adjusting anteversion and inclination angle of the acetabulum fragment; (**b**) biomechanical optimization; (**c**) the pre-operative planning output

1.2 Materials and Methods

1.2.1 System Overview

The workflow of the computer assisted planning of PAO with biomechanical optimization is shown in Fig. 1.1 The input is 3D surface models of pelvis, femur and their respective cartilages generated from pre-operatively acquired CT arthrography data using a commercially available segmentation program (AMIRA, Visualization Sciences Group, Burlington, MA). The system starts with a fully automatic detection of the acetabular rim, which allows for quantifying the acetabular morphology with parameters such as acetabular version, inclination, LCE angle, femoral head extrusion index (EI), femoral head coverage ratio (CO) for a computer assisted diagnosis [9]. Afterwards, the system offers the possibility to perform a virtual osteotomy (see Fig. 1.1a(1)) and reorient the acetabular fragment. During the acetabulum fragment reorientation, acetabular morphological parameters are computed in real-time (see Fig. 1.1a(2)). In order to estimate the optimal acetabulum reorientation for planning PAO, our system is extended with a patient-specific finite element prediction of cartilage contact stress change before and after PAO reorientation planning. An optimal position of the acetabulum can be achieved, which maximizes contact area and at the same time minimizes peak contact pressure in pelvic cartilage (see Fig. 1.1b).

1.2.2 *Biomechanical Model of Hip Joint*

1.2.2.1 Cartilage Models (Constant Thickness Cartilage vs. Patient-Specific Cartilage)

In the literature, both constant thickness cartilage models and patient-specific cartilage models have been suggested. For instance, Zou et al. [8] created the cartilage layer on acetabular surface by expanding a constant thickness of 1.8 mm. Harris et al. [11] introduced a CT arthrography protocol for excellent visualization patient-specific cartilage geometry. In our study, the patient-specific cartilage models were generated from the CT arthrography data for the subjects with traditional acetabular dysplasia [12]. The constant thickness cartilage models were generated by expanding a constant thickness using 3D dilation operation on articular surface.

1.2.2.2 Mesh Generation

Bone and cartilage surface models of the reoriented hip joints were imported into ScanIP software (Simpleware Ltd, Exeter, UK) as shown in Fig. 1.2a, c. Surfaces were discretized using tetrahedral elements (Fig. 1.2b, d). Since the primary concern was focused on the joint contact, a finer mesh was employed for the cartilage than for the bone. Refined tetrahedral meshes were constructed for the cartilage models ($\sim$106,672 elements for femoral cartilage model, and $\sim$55,476 elements for pelvic cartilage model) using ScanFE module (Simpleware Ltd, Exeter, UK). Cortical bone surfaces were discretized using coarse tetrahedral elements ($\sim$99,023 elements for femoral model, and $\sim$128,745 elements for pelvic model). Trabecular bone was not included in the models, as it only has a minor effect on the predictions of contact stress as reported in [13].

1.2.2.3 Material Property

Pelvic and femoral cartilages were modeled as homogeneous, isotropic, and linearly elastic material with Young's Modulus $E = 15$ MPa and Poisson's ratio $\nu = 0.45$ as reported in [8]. Cortical bone of pelvis and femur were modeled as homogeneous, isotropic material with elastic modulus $E = 17$ GPa and Poisson's ratio $\nu = 0.3$ as suggested in [8].

1.2.2.4 Boundary Conditions and Loading

Tied and sliding contact constraints were used in Abaqus/CAE 6.10 (Dassault Systèmes Simulia Corp, USA) to define the cartilage-to-bone and cartilage-to-cartilage interfaces, respectively. It has been reported in [11] that the friction coefficient between articular cartilage surfaces was very low (0.01–0.02 in the

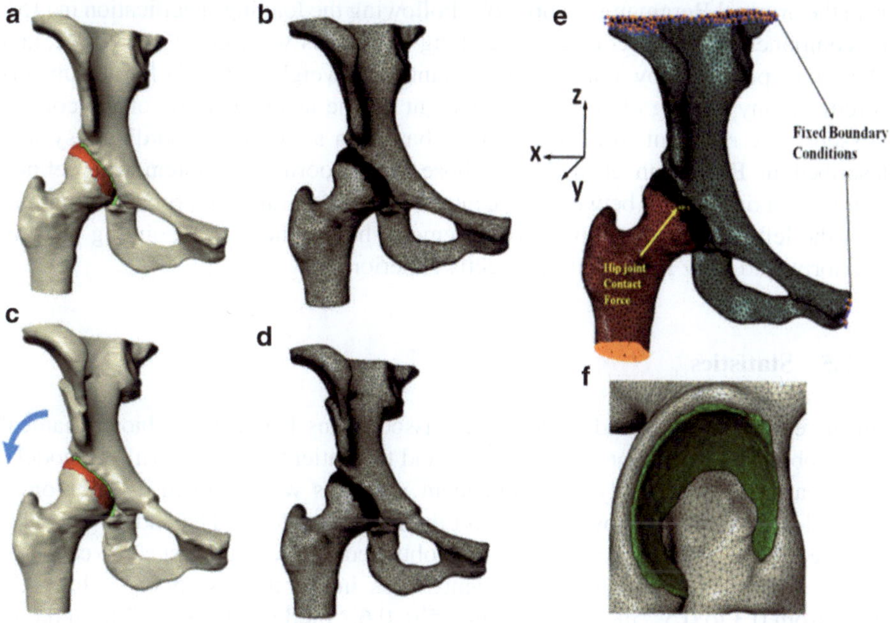

Fig. 1.2 Biomechanical simulation of contact pressure on acetabular cartilage. (**a**) surface models of a dysplastic hip; (**b**) volume meshes of a dysplastic hip; (**c**) surface models for a planned situation after acetabulum fragment reorientation; (**d**) volume meshes for the planned situation; (**e**) boundary conditions and loading for biomechanical simulation; (**f**) coarse meshes for bone models, and refined meshes for cartilages

presence of synovial fluid) [14]. Therefore, it is reasonable to neglect frictional shear stresses between contacting articular surfaces. The loading and boundary conditions used in this paper resembled those used by Phillips et al. [15] (Fig. 1.2e). The top surface of pelvis and pubic areas were fixed, and the distal end of the femur was constrained to prevent displacement in the body x and y directions while being free in vertical z direction (Fig. 1.2e). The center of femoral head derived from a least-squares sphere fitting was selected to be the reference node. The nodes of femoral head surface were constrained by the reference node via kinematic coupling. The fixed boundary condition model was then subjected to a loading condition as published in [16], representing a single leg stance situation with the resultant hip joint contact force acting at the reference node. Although CT scan was performed in the supine position and the loading condition of our biomechanical simulation is based on one-leg stance situation [16], previous work [17] has shown that there was no significant difference between the contact pressure in the one-leg stance reference frame and those in the supine reference frame. In addition, as pointed out by Armiger et al. [7], it is not an infrequent clinical practice to use models derived from the supine frame to do biomechanical simulation of the standing frame. Therefore we believe that our model makes good use of valuable, available data from the original Bergmann's work [16]. Following the loading specification in [15], the components of joint contact force along three axes were given as 195, 92, and 1490 N, respectively, by assuming a constant body weight of 650 N for all subjects to remove any scaling effect of body weight on the absolute value of the contact pressure. The resultant force was applied based on anatomical coordinate system described in Bergmann et al. [16], whose local coordinate system was defined with the x axis running between the centers of the femoral heads (positive running from the left femoral head to the right femoral head), the y axis pointing directly anteriorly, and the z axis pointing directly superiorly.

1.2.2.5 Statistics

Linear regression was used to determine associations between the biomechanical results obtained by the constant thickness and the patient-specific cartilage models. For linear regression analysis, independent variables were defined as the biomechanical results obtained by the constant thickness cartilages. Dependent variables were defined as the biomechanical results obtained by the patient-specific cartilage models. Pearson's correlation coefficient r was interpreted as "poor" below 0.3, "fair" from 0.3 to 0.5, "moderate" from 0.5 to 0.6, "moderately strong" from 0.6 to 0.8, and "very strong" from 0.8 to 1.0. Significance level was defined as $p<0.05$.

1.2.3 Study Design and Results

Two studies were designed and conducted on CT arthrography data of three patients with acetabular dysplasia [12]. The first study is designed to estimate the optimal orientation after a computer assisted planning of PAO based on a 3D FE simulation. Specifically, the acetabulum fragment is virtually rotated about the y axis (Fig. 1.2e) in 5° increment onto the APP from the original acetabulum inclination angle towards lateral direction (see Fig. 1.2c). The predicted peak contact pressure and total contact area are directly extracted from the output of Abaqus. We then compared quantitatively the peak contact pressure and contact area on acetabulum cartilage in different acetabulum position and estimated optimal orientation in static one-leg stance loading scenario.

Figure 1.3 shows how contact pressure distribution of the pelvic cartilage changed for a dysplastic hip when LCE angle was increased. The contact area originally focused on the anterosuperior region and almost no contact area was in the anterior and posterior regions. When the LCE angle was increased, the contact area shifted from lateral region towards the medial region. Figure 1.4a, b shows peak contact pressures and contact area at different LCE angles, respectively. An optimal acetabulum fragment reposition with minimum peak contact pressure and maximum contact area was achieved for three dysplastic hips. More importantly, for each hip, both the minimal peak contact pressure and the maximum contact

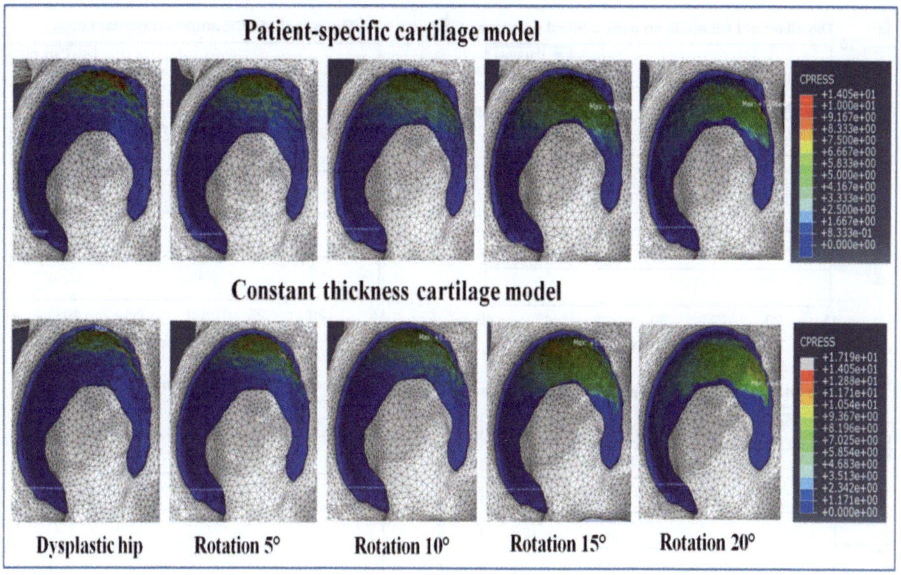

Fig. 1.3 Contact pressure distribution obtained by using two different cartilage models at different acetabular reorientation position

area were achieved at the same acetabulum fragment reposition. A large rotation of acetabulum does not guarantee low peak contact pressure and large contact area.

The second study is designed to investigate the effect of these two different cartilage models on the biomechanical optimization results. Figure 1.3 qualitatively shows that contact pressure distribution obtained by using two different cartilage models was similar. Regression analysis quantitatively shows that the results obtained by the constant thickness cartilage models are significantly correlated with those obtained by using the patient-specific cartilage models. Specifically, a very strong correlation is between the peak contact pressure obtained by two different cartilage models ($r = 0.93 > 0.8$, $p = 0.013 < 0.05$) (see Fig. 1.4c), and a moderate strong correlation is between the total contact area obtained by two different cartilage models ($r = 0.72 \in [0.6, 0.8]$, $p = 0.16$). Furthermore, biomechanical optimization-based planning of PAO using these two different cartilage models achieved the same optimal orientations (see Fig. 1.4d, e).

1.3 Discussion and Conclusions

In this paper, we developed a patient-specific FE model for optimization of the acetabulum reorientation based on our previously developed computer assisted planning system. Because articular cartilage is a key component to affect biomechanical

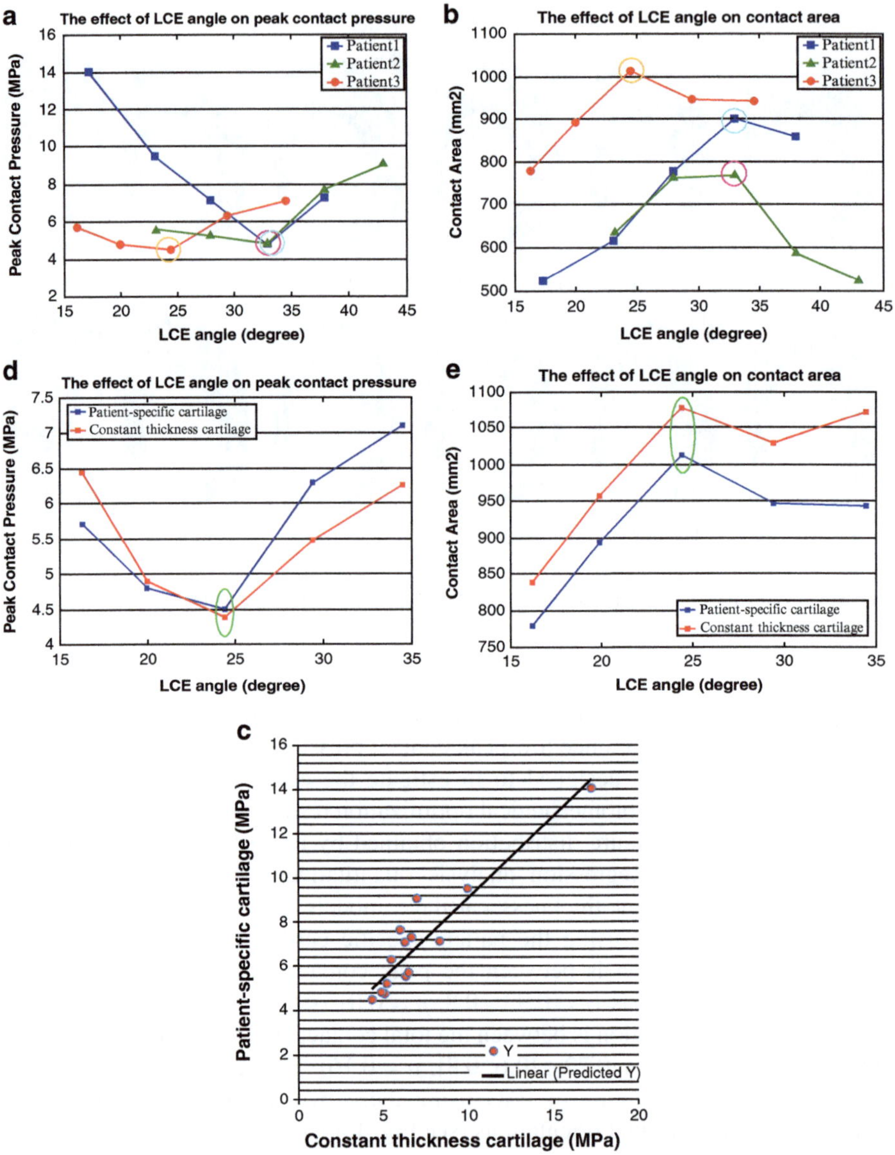

Fig. 1.4 (**a**) Effect of LCE angle on hip joint peak contact pressure. *Circled points* represent the lowest pressures for each subject. (**b**) Effect of LCE angle on hip joint contact area. *Circled points* indicate the largest contact areas for each subject. (**c**) Scatter plot of peak contact pressure obtained by constant thickness cartilage models against those obtained by patient-specific cartilage models. The *solid bar* represents regression line. (**d**) Effect of LCE angle on peak contact pressures predicted by two different cartilage models. (**e**) Effect of LCE angle on contact areas predicted by two different cartilage models

optimization results, we further investigated the influence of patient-specific and constant thickness cartilage models on the biomechanical optimization results. Regression analysis showed that the results obtained by the constant thickness cartilage models are significantly correlated with those obtained by using the patient-specific cartilage models. Furthermore, biomechanical optimization-based planning of PAO using these two different cartilage models achieved the same optimal orientations.

It is known that the purpose of PAO is to increase acetabular coverage of the femoral head and thereby decrease contact pressure over the cartilage surface [18], but a limitation common to all these previously introduced methods [3–5] is that they only provide a pure morphology-based assessment and/or planning without considering the biomechanical effect of the reorientation planning. This has motived the recent introduction of various systems based on biomechanical optimization [2, 6–8]. The BGS introduced by Armand et al. [2] performed DEA to estimate the contact pressure on a patient-specific model which is a computationally efficient method for modeling of cartilage stress while neglecting underlying bone stress. Different from the method of Armand et al. [2], we developed a 3D patient-specific FE model for biomechanical analysis derived from our computer assisted planning system (see Fig. 1.1a(1)). The result of our FE simulation study shows that our computer assisted planning system with biomechanical optimization indeed reduces contact pressures and at the same time increases contact areas, which is consistent with the study results reported by Armand et al. [2].

Compared to the results reported by Zou et al. [8], who developed a 3D FE simulation of PAO in order to find optimal reorientation position by minimizing peak contact pressure and at the same time maximizing contact area of the cartilage surfaces, our results are also consistent with theirs. Both studies have proved that 3D FE model is an efficient tool to predict cartilage contact stress change before and after PAO reorientation planning [8].

Another contribution of our paper is the investigation of the effect of these two different cartilage models on the biomechanical results. In the literature, little is known about how different cartilage models used in the FE simulation may further affect the biomechanical optimization-based PAO planning. Niknafs et al. [17] investigated the effects of four different cartilage thickness profiles (one uniform and three non-uniform) and two sets of compressive properties on optimal alignment planning for PAO based on the BGS introduced by Armand et al. [2]. The result of their study shows that the predicted optimal alignment of the acetabulum was not significantly sensitive to the choice of cartilage thickness distribution over the acetabulum. Our experimental results found that there was no statistically significant difference for two different cartilage models, which is consistent with the findings of Niknafs et al. [17]. Another study was introduced by Anderson et al. [19], who analyzed 3D FE models of hip cartilage mechanics with simplified geometrical model to investigate the effects on predictions of cartilage stress. Their study showed that pressures were slightly different and pressure distribution was similar for patient-specific and constant thickness models in normal hips. However, their study did not investigate the effect of different cartilage models on the biomechanical optimization results during virtual reorientation of dysplastic hips.

It is worth to mention the limitations of the present method. The main limitation is that the acetabular labrum was neglected to be included in our FE models of dysplastic hips. The role of the labrum during load distribution has been debated. Based on the measurements from pressure-sensitive film, Konrath et al. [20] concluded that there were slight changes in contact area, mean pressure, or maximum pressure in the anterior or superior acetabulum. The only significant change was a decrease in the maximum pressure in the posterior aspect of the acetabulum. In contrast, an in vitro study by Ferguson et al. [21] found that the labrum has an influence on intra-articular fluid pressurization and cartilage layer consolidation in the hip joint. A recent study by Henak et al. [22] found that the labrum supported less than 3 % of the total load across the joint in normal hips. More recently, Henak et al. [12] found that the labrum in dysplastic hips has a far more significant role in hip mechanics than it does in normal hips. Their study demonstrated that cartilage contact stresses in dysplastic hips are not increased significantly compared to normal hips because the labrum supports a large percentage of the load transferred across the joint. Therefore inclusion of labrum geometry is necessary for more realistic and accurate FE model.

Another limitation is that a fixed body weight of 650 N derived from Bergmann et al. [16] was applied to all three dysplastic hips for 3D FE simulation, which is not patient-specific. The argument why we adopted such a strategy is that we are aiming to compare the relative change of contact pressure before and after PAO reorientation planning. Thus, it makes sense to use a constant loading, which was originally proposed by Zou et al. [8]. The last limitation of our study is also related with the loading conditions for our 3D FE prediction. In our study, data reported by Bergmann et al. [16] was used as in vivo contact force for FE simulation, which was obtained from patients underwent total hip arthroplasty (THA). This may be overcome by using in vivo contact force of dysplastic hip derived from force sensor, which is beyond the scope of this paper.

In conclusion, this study suggested that our computer assisted planning with FE modeling can be a promising PAO planning tool even when it is not feasible to obtain CT arthrography data in clinical routine.

Acknowledgements This work was supported by the open source dysplastic hips image data from the University of Utah (http://mrl.sci.utah.edu/software/dysplastic-hips-image-data/) and partially supported by the Japanese-Swiss Science and Technology Cooperation Program.

References

1. S.B. Murphy, M.B. Millis, J.E. Hall, Surgical correction of acetabular dysplasia in the adult. A Boston experience. Clin. Orthop. Relat. Res. **363**, 38–44 (1999)
2. M. Armand, J.V.S. Lepistö, A.C. Merkle et al., Computer-aided orthopedic surgery with near-real-time biomechanical feedback. J. Hopkins APL Tech. Dig. **25**(3), 242–252 (2004)

3. J.C. Clohisy, J.C. Carlisle, R. Trousdale et al., Radiographic evaluation of the hip has limited reliability. Clin. Orthop. Relat. Res. **467**(3), 666–675 (2009)
4. K. Klaue, A. Wallin, R. Ganz, CT evaluation of coverage and congruency of the hip prior to osteotomy. Clin. Orthop. Relat. Res. **232**, 15–25 (1988)
5. W. Dandachli, V. Kannan, R. Richards et al., Analysis of cover of the femoral head in normal and dysplastic hips: new CT-based technique. J. Bone Joint Surg. (Br.) **90**(11), 1428–1434 (2008)
6. X. Zhao, E. Chosa, K. Totoribe, Effect of periacetabular osteotomy for acetabular dysplasia clarified by three-dimensional finite element analysis. J. Orthop. Sci. **15**(5), 632–640 (2010)
7. R.S. Armiger, M. Armand, K. Tallroth et al., Three-dimensional mechanical evaluation of joint contact pressure in 12 periacetabular osteotomy patients with 10-year follow-up. Acta Orthop. **80**(2), 155–161 (2009)
8. Z. Zou, A. Chavez-Arreola, P. Mandal, T.N. Board, T. Alonso-Rasgado, Optimization of the position of the acetabulum in a Ganz periacetabular osteotomy by finite element analysis. J. Orthop. Res. **31**(3), 472–479 (2012)
9. L. Liu, T.M. Ecker, S. Schumann et al., Computer assisted planning and navigation of periacetabular osteotomy with range of motion optimization, in Medical Image Computing and Computer-Assisted Intervention (MICCAI 2014) (Springer International Publishing, 2014), pp. 643−650
10. G. Zheng, A. Marx, U. Langlotz et al., A hybrid CT free navigation system for total hip arthroplasty. Comput. Aided Surg. **7**(3), 129–145 (2002)
11. M.D. Harris, A.E. Anderson, C.R. Henak et al., Finite element prediction of cartilage contact stresses in normal human hips. J. Orthop. Res. **30**(7), 1133–1139 (2012)
12. C.R. Henak, C.L. Abraham, A.E. Anderson et al., Patient-specific analysis of cartilage and labrum mechanics in human hips with acetabular dysplasia. Osteoarthr. Cartil. **22**(2), 210–221 (2014)
13. A.E. Anderson, B.J. Ellis, S.A. Maas et al., Validation of finite element predictions of cartilage contact pressure in the human hip joint. J. Biomech. Eng. **130**(5), 051008 (2008)
14. M. Caligaris, G.A. Ateshian, Effects of sustained interstitial fluid pressurization under migrating contact area, and boundary lubrication by synovial fluid, on cartilage friction. Osteoarthr. Cartil. **16**(10), 1220–1227 (2008)
15. A.T.M. Phillips, P. Pankaj, C.R. Howie et al., Finite element modelling of the pelvis: inclusion of muscular and ligamentous boundary conditions. Med. Eng. Phys. **29**, 739–748 (2007)
16. G. Bergmann, G. Deuretzabacher, M. Heller et al, Hip contact forces and gait patterns from routine activities. J. Biomech. **34,** 859–871 (2001). Including the HIP98 CD, 3-9807848-0-0
17. N. Niknafs, R.J. Murphy, R.S. Armiger et al., Biomechanical factors in planning of periacetabular osteotomy. Biomechanics **1**(20), 1–10 (2013)
18. A. Iglic, V.K. Iglic, V. Antolic et al., Effect of the periacetabular osteotomy on the stress on the human hip joint articular surface. IEEE Trans. Rehabil. Eng **1**(4), 207–212 (1993)
19. A.E. Anderson, B.J. Ellis, S.A. Maas et al., Effects of idealized joint geometry on finite element predictions of cartilage contact stresses in the hip. J. Biomech. **43**(7), 1351–1357 (2010)
20. G.A. Konrath, A.J. Hamel, S.A. Olson et al., The role of the acetabular labrum and the transverse acetabular ligament in load transmission in the hip. J. Bone Joint Surg. Am. **80**(12), 1781–1788 (1998)
21. S.J. Ferguson, J.T. Bryant, R. Ganz et al., An in vitro investigation of the acetabular labral seal in hip joint mechanics. J. Biomech. **36**, 171–178 (2003)
22. C.R. Henak, B.J. Ellis, M.D. Harris et al., Role of the acetabular labrum in load support across the hip joint. J. Biomech. **44**(12), 2201–2206 (2011)

Chapter 2
Role of Ligaments in the Knee Joint Kinematic Behavior: Development and Validation of a Finite Element Model

F. Germain, P.Y. Rohan, G. Rochcongar, P. Rouch, P. Thoreux, H. Pillet, and W. Skalli

2.1 Introduction

The management of knee instability is a complex problem in orthopedic surgery. It usually involves the tear or the rupture of a cruciate or collateral ligament of the articulation and, in many cases, necessitates surgical operation and ligament reconstructions. The rupture of the ACL, in particular, is one of the most frequently occurring ligament injuries affecting about one person in 3000 every year and this trend has been constantly increasing with the rise of participation in sports in the general population [1]. Knee ligamentoplasty has become an issue of high clinical interest.

The current standard of care is based on ligament reconstruction by autografts from tendon tissues, allograft, or ligament substitutes. The assessment, however, of the long-term performance of the surgery is complicated by the complex interaction between the graft parameters (pretensioning, fixation method, etc.), knee anatomy, and the mechanical interaction between the graft and the passive anatomical structures. To this day, no diagnostic tool is available in clinics to quantitatively evaluate and predict the impact of surgery on knee kinematics. This highlights the need for the development of tools to investigate the contribution on knee joint mobility of the ligaments in the normal knee and to establish proper treatment strategies.

In that respect, many studies have been performed both in vitro [2–5] and in vivo [6–9] to study knee kinematics, some of them focusing on influence of the main knee ligaments [10, 11]. The results varied greatly across studies highlighting the high inter-individual variability.

F. Germain • P.Y. Rohan (✉) • G. Rochcongar • P. Rouch • P. Thoreux • H. Pillet • W. Skalli
LBM/Institut de Biomécanique Humaine Georges Charpak, Arts et Metiers ParisTech, 151 bvd de l'Hôpital, 75013 Paris, France
e-mail: pierreyves.rohan@live.fr

G. Joldes et al. (eds.), *Computational Biomechanics for Medicine*,
DOI 10.1007/978-3-319-28329-6_2

Several finite element (FE) knee models have also been developed in an effort to comprehensively investigate various aspects of knee mechanics including contact pressure under various loads [12–18], ligament stress [19–21], or the dynamic behavior [14, 22, 23]. A few of them studied knee kinematics [12, 14, 24, 25]. The high numerical costs associated with the computation of the biomechanical response of the knee joint and the full validation of these models against tibio-femoral and tibio-patellar kinematic data, however, are still a main issue [14], particularly for patellofemoral motion because of uncertainty in patellar tracking.

A new experimental setup was recently developed combining 3D reconstruction imaging with the use of a motion capture system for accurate analysis of knee joint kinematics at our laboratory [2]. The aim of the present study was to develop an FE model capable of reproducing the kinematic of the knee in flexion/extension and, building upon the work of Azmy, to validate it with in vitro experimentations for investigating the impact of ligament properties on the knee kinematics.

2.2 Materials and Methods

2.2.1 Finite Element Model

Geometry The model developed in this study is an adaptation of the one proposed by [26]. The 3D geometry of the bony structures was acquired using a sensor pen (Fastrak system, Polhemus, Olchester, USA), and was carried out on a left 63-year-old female lower limb.

Elements and Meshing The FE mesh, represented in Fig. 2.1, consists of the distal and proximal part of the femur and tibia, respectively, the patella, the quadriceps tendon, the cartilage, and knee ligaments. Bones and cartilage were meshed with low-order (four-nodes) shell elements, the quadriceps tendon with low-order membrane elements, and the ligaments with tension-only cables.

The different ligament bundles were represented: (1) two bundles for each cruciate ligament (antéro-médial (AM) and postéro-lateral (PL) bundles for the Anterior cruciate ligament (ACL); antéro-latéral (AL) and postéro-médial (PM) bundles for the Posterior cruciate ligament (PCL)), (2) four bundles for the Collateral Ligaments (CoL) (three for the medial collateral ligament (MCL) and one for the lateral collateral ligament (LCL)), (3) five bundles for the patellar ligament, and (4) seven for the articular capsule.

Material Properties The material properties used for each component are summarized in Tables 2.1 and 2.2. Bones and cartilage were defined as linear elastic isotropic material in accordance with the literature [27, 28]. The regions covered with cartilage were modeled as a homogeneous bone cartilage material, with average properties. A thin strip of elements between the bones and cartilage

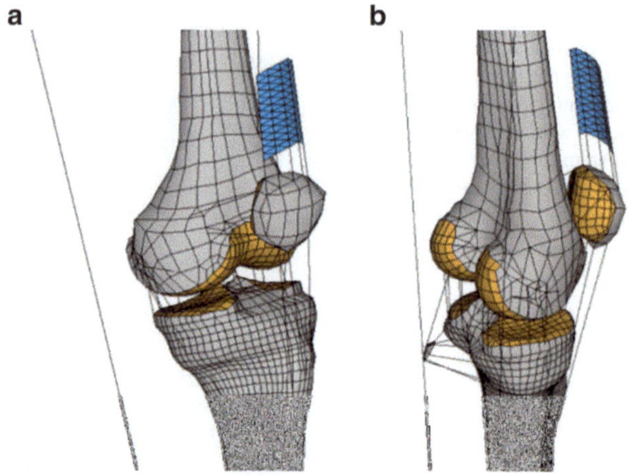

Fig. 2.1 FE knee model. (**a**) Isometric view (*front*); (**b**) isomeric view (*back*). The femur and tibia were limited to their respective distal and proximal parts

Table 2.1 Material properties of bones and cartilage

	E (MPa)	ν
Cortical bone and patella	12,000	0.3
Bone/cartilage inter	2000	0.4
Bone/cartilage	250	0.4
Tendon	90	0.4

Table 2.2 Material properties of the different ligament bundles

	K (N/mm)	Initial strain (%)
Anterior cruciate ligament (ACL)	75	5
Posterior cruciate ligament (PCL)	75	−3
MCL	70	0
LCL	20	0

regions were also defined with intermediate properties in order to avoid important mechanical discontinuity.

The stiffness of the different ligament bundles was chosen in accordance with the experimental data reported in literature [29–34]. Negative initial strain for PCL indicates a slack condition.

Contact Four frictionless surface-to-surface contact zones were created: femur (lateral)/tibia (lateral), femur (medial)/tibia (medial), femur/patella, and femur/quadriceps tendon. The interactions between the ligaments and bones represented in the model are not modeled in the present study.

Boundary Conditions The experimental boundary conditions [2] were imposed: the femur was fixed, the rope and pulley system was represented as two cables driven by temperature, and a set of 130 consecutive displacements of 0.5 mm were applied to the quadriceps tendon. This allowed to generate forces in the cables that were always directed toward the center of the femoral head.

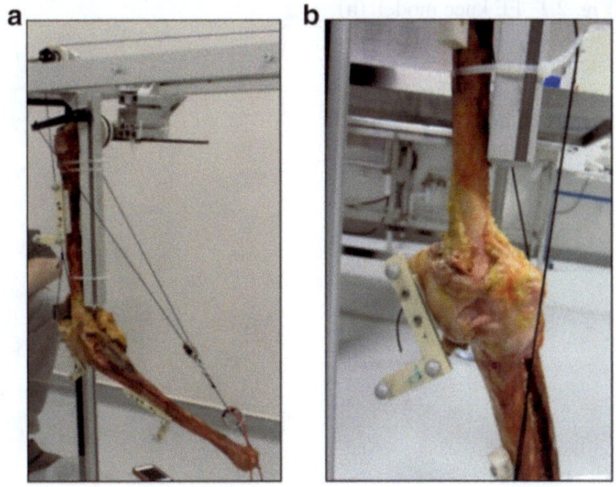

Fig. 2.2 Validated experimental setup used for testing the lower limbs in flexion/extension. (**a**) Whole setup with the rope and pulley system for applying the flexion. (**b**) Position of the tripods

Solution The large deformation static response was computed using an implicit solver in ANSYS.

2.2.2 *Model Evaluation*

Sample Preparation Twenty-three fresh frozen lower limbs were used. The subjects from whom they were harvested were 47–97 years old. Absence of osteoarthritis and ligament laxity was checked. Specimens were disarticulated at the head of the femur and at the distal epiphysis of the tibia. All the soft tissues were removed except for the articular capsule of the knee, the quadriceps tendon, the CoL of the knee (medial and lateral), the patellar ligament, and the proximal and distal ligaments between the tibia and the fibula. The samples were frozen at −20 °C then thawed at room temperature 12 h before experimentation.

Test Bench The validated test bench, represented in Fig. 2.2, was adapted from [2]. The femur was fixed and the tibia was free. Tension was applied to the quadriceps tendon using a weight of 9.8 N. A motor was connected to the tibial pilon by a rope and was used to perform the flexion movement. A progressive displacement was applied at the centroid of the malleolus and directed toward the center of the femoral head.

Movement Tracking Tripods were fixed on the femoral diaphysis, the proximal epiphysis, and the patella. During the cycles of flexion/extension, the position of the bony segments was tracked with a Polaris optoelectronic system (NDI, Waterloo, Ontario, Canada).

Anatomical Frames The position of the tripods in their respective anatomical frames was calculated from 3D reconstruction made with the EOS® bi-planar

Table 2.3 Stiffness and initial strain chosen for the sensitivity cases on the ACL

	K (N/mm)			Initial strain (%)		
	Min	Mean	Max	Min (%)	Mean (%)	Max (%)
ACL bundle	40	75	150	0	3	6
PCL bundle	40	75	150	3	0	−4
MCL	70	100	120	0	2	4
LCL	20	60	120	0	2	4

X-ray system. The experimental setup was put in the EOS cabin and bi-planar radiographies were acquired. A geometric reconstruction of the bony structures was performed and used for the calculation of the anatomical frames.

Movement Analysis The variation of relative position tibia/femur and patella/femur was extrapolated using a MATLAB® routine. The rotations were calculated on mobile axis, with the following sequence (center of rotations): $zy'x''$ (y' denoting the mobile y axis after the first rotation about z, and x'' denoting the mobile x axis after the first two rotations).

2.2.3 Sensitivity Analysis

A sensitivity analysis was conducted to investigate the impact of ligaments on the knee kinematics. The stiffness and initial strain of one or several bundles of ligament were modified and the impact in the tibia position was computed. Ten cases were considered. These are summarized in Table 2.3. Stiffness values were defined based on data reported in the literature [29–34]. Initial strain values were chosen so that strains remained below the physiological limit whilst being sufficiently different from one another.

2.3 Results

2.3.1 Experimental/Numerical Comparison

Tibial Kinematics Both the experimental corridor of the tibial kinematics with respect to the femur and the numerical results computed with our model are shown in Fig. 2.3. Our results show that the angular position of the tibia of the finite element model is in the physiological corridor. During flexion, the tibial movement predicted by the FE model does an internal rotation (Fig. 2.3) of up to −11.9° at 70° flexion (while the in vitro experiment is in the corridor [−18.3°; −7.2°]). Our results also

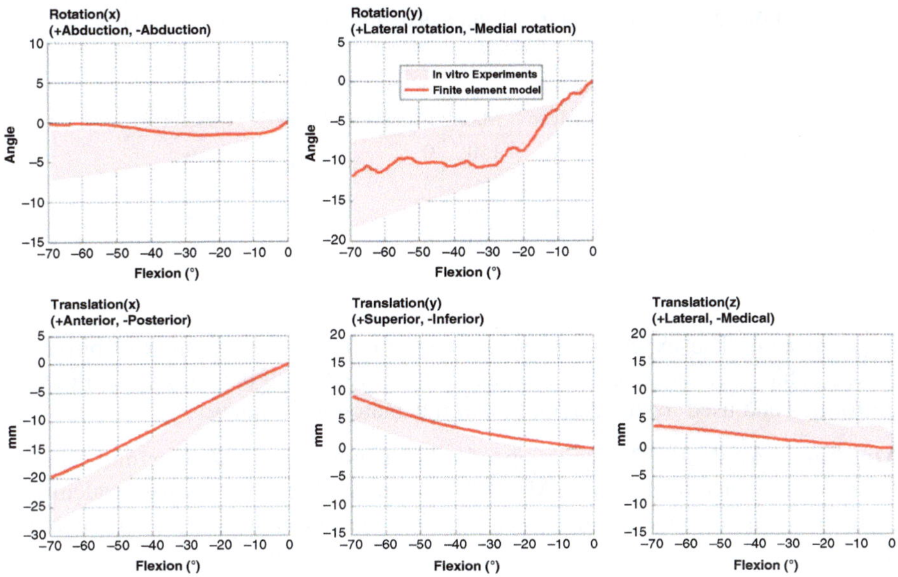

Fig. 2.3 Tibial kinematics (with respect to the femur) during the range of motion

show a slight adduction: down to −1.71° at 20° flexion and then back to −0.26° at 70° of flexion ([−7.1°; −0.9°] for the in vitro experiments).

Patellar Kinematics Both the experimental corridor of the patellar kinematics with respect to the femur and the numerical results computed with our model are shown in Fig. 2.4. The results obtained using the finite element model are within the physiological corridor of the *x*, *y*, and *z* rotations and the *z* translation.

2.3.2 Influence on the Femoro-Tibial Kinematics

The influence of the ligament parameters on the femoro-tibial kinematics is reported in Table 2.4, mean value and standard deviation of tibia position at 60° flexion for each case. Concerning the Rx rotation (abduction/adduction), the ligament that had the most influence was the ACL, followed by the LLE, the PCL, and the MCL holding a smaller influence (SD < 0.35°). The Ry rotation is mainly influenced by the PCL and the LCL, the PL bundle of ACL holding a smaller influence (SD < 1.01°). The Tx translation was influenced mainly by the ACL and the PCL when both bundles are modified at the same time. The Ty translation was influenced mainly by the AM bundle of the MCL and the cruciate ligaments when the four ligaments are modified at the same time. The Tz translation is influenced mainly by the ACL, the AL bundle of the PCL, and the LCL.

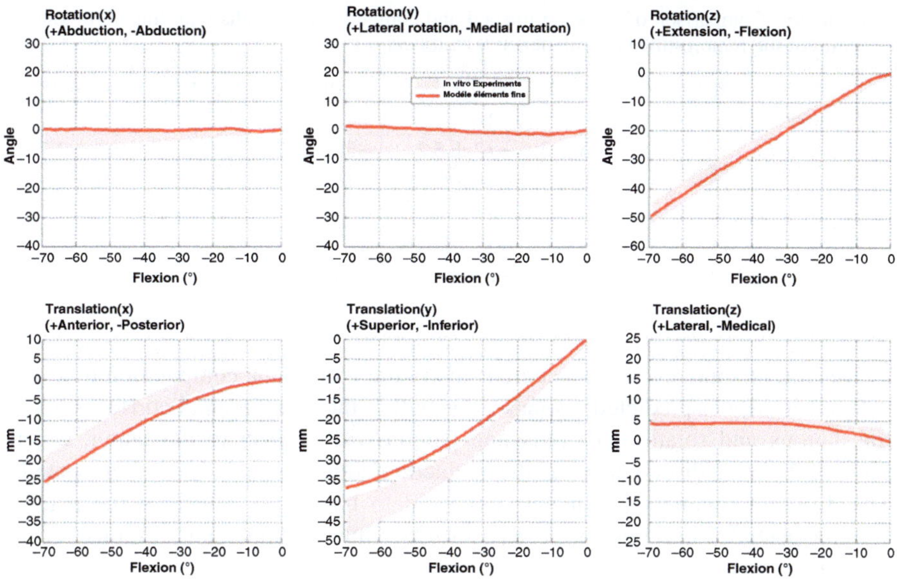

Fig. 2.4 Patellar kinematics (with respect to the femur)

Table 2.4 Mean value and standard deviation of tibia position at 60° flexion for each sensitivity case (the ligament column shows which ligament was modified)

Ligament		Rx (°)	Ry (°)	Tx (mm)	Ty (mm)	Tz (mm)
ACL (AM)	Mean	−1.2	−13.9	−17.3	5.9	3.0
	SD	1.0	0.5	0.5	0.5	0.2
ACL (PL)	Mean	−0.5	−15.5	−16.8	5.0	3.0
	SD	0.6	1.0	0.9	0.2	0.2
ACL (all)	*Mean*	*−1.3*	*−14.2*	*−17.2*	*5.9*	*2.9*
	SD	*1.0*	*0.3*	*0.4*	*0.6*	*0.2*
PCL (AL)	Mean	0.0	−15.6	−17.9	5.1	3.1
	SD	0.1	1.8	0.3	0.1	0.1
PCL (PM)	Mean	0.1	−15.2	−17.7	5.1	3.2
	SD	0.3	1.4	0.2	0.3	0.0
PCL (all)	*Mean*	*0.3*	*−14.8*	*−17.2*	*5.0*	*3.2*
	SD	*0.3*	*3.0*	*0.8*	*0.4*	*0.1*
MCL (all)	Mean	0.2	−14.2	−17.7	5.2	3.2
	SD	0.2	0.3	0.1	0.1	0.1
LCL	Mean	0.2	−14.6	−17.6	5.1	3.2
	SD	0.7	1.7	0.2	0.1	0.2
CoL (all)	*Mean*	*0.2*	*−14.3*	*−17.*	*5.1*	*3.0*
	SD	*0.7*	*1.3*	*0.1*	*0.1*	*0.1*

Table 2.5 Comparison of the experimental and computational results with literature: kinematics of the femoro-tibial articulation at 60° flexion

Study	Specimens	Rx (°)	Ry (°)	Tx (mm)	Ty (mm)	Tz (mm)
FE model	–	0	−11.5	−17.2	7.5	4.9
In vitro exp	23	-3 ± 3	-12 ± 5.2	-22.5 ± 2.5	5.5 ± 2.5	3.3 ± 2.7
[35]	15	-3 ± 3	-16.5 ± 5	-16.7 ± 3.7	7.5 ± 3	4 ± 1.5
[5]	8	–	-11 ± 4	–	–	–
[36]	15	–	-17.1 ± 1.8	–	–	–
[3]	13	–	-6.5 ± 3	-14 ± 5	–	–

Despite the numerous contact elements in the model, the computation time was below 1 h. Model-predicted response was within the experimental corridors for all translations and rotations of tibia and patella with respect to the femur, except for a ±2° max discrepancy in the abduction/adduction rotation of the patella. The different ligament parameters appeared to have little effect on the patellar anterior/posterior and inferior/posterior shifts while the kinematic response of the tibial anterior/posterior shift was more sensitive.

2.4 Discussion

Very few FE models reported in the literature feature extensive validation against both tibio-femoral and tibio-patellar kinematic data. These are paramount to comprehensively assess the biomechanical role of the knee joint and to investigate various aspects of knee instability such as knee ligament injuries, ligament replacement, and ACL graft design.

In this contribution, an FE of the knee joint capable of reproducing the kinematics of the knee in flexion/extension for the investigation of the role of ligaments in the joint kinematic behavior was created and validated against the experimental data of 23 cadavers. Special attention was paid to keep the numerical cost as low as possible. Low-order shell elements and cable elements were therefore used to discretize the components of the geometric model.

For both the tibial and patellar kinematics, the results computed with our model are inside or at the limit of the experimental corridors. Both the experimental and computational results obtained for the femoro-tibial and the femoro-patellar kinematics are comparable to the literature. Tables 2.5 and 2.6 summarize and compare the kinematic values reported in this study with the main values reported in the literature.

As for model evaluation, Kiapour et al. [14] presented extensive validation using 16 lower limbs and investigation of several parameters. However, patellofemoral kinematics was not evaluated.

Table 2.6 Comparison of the experimental and computational results with literature: kinematics of the femoro-patellar articulation at 60° flexion

Study	Specimens	Rx (°)	Ry (°)	Tx (mm)	Ty (mm)	Tz (mm)
FE model	–	0	0.5	–41.5	–20	–34
In vitro exp	23	-3 ± 3	-3.5 ± 4.5	-41.5 ± 3.5	-17 ± 3	-41 ± 4
[2]	8	-3 ± 3	0.5 ± 7.5	-38 ± 4	-14 ± 4	-40 ± 8
[5]	8	1.5 ± 3	3 ± 5.5	–	–	–
[37]	6	1 ± 5.5	-1 ± 10	–	–	–
[4]	7	0 ± 4	-2 ± 5	–	–	–

The sensitivity analysis conducted on the main ligaments of the tibio-femoral articulation highlighted the importance of distinguishing the two bundles of the ACL and the PCL, since they do not impact the femoro-tibial kinematics in the same way: the AM bundle of the ACL exerts more than twice as much influence on Ry and Tx, but less than twice as much on Rx and Ty as compared with the PL bundle. Likewise, modifying both bundles of the PCL significantly increases the influence on Rx, Ry, and Tx, but doesn't have much effect on Ty and Tz.

Two main limitations of our model are the omission of (1) the meniscus, the popliteal tendon, and the popliteo-fibular ligament, in order to keep the computation time as low as possible. This might explain the small experimental–numerical kinematic discrepancies and could be taken into account to improve the realism of the model. The simplification of the ligament insertion and the absence of ligament/ligament contact could also explain the fact that, in the simulation results, the ligament effect remains small. (2) The geometry of the finite element model is not personalized. Yet the geometry has a strong impact on the femoro-tibial and the femoro-patellar kinematics for at least two reasons: it affects the morphology of the contact surface, and it can determine the position of anatomical landmarks from which the boundary conditions are defined (center of the femoral head, center of the malleoli).

Nonetheless, the model appears consistent with in vitro kinematics and the computational cost remains reasonable (1 h). From a clinical perspective, the developed knee model constitutes a valuable tool for exploring new treatment strategies, such as the development of ACL substitutes for ligament reconstructions where it can be used to test different designs and evaluate the restoration of knee kinematics right after implantation.

2.5 Conclusions

An FE of the knee joint capable of reproducing the kinematic of the knee in flexion/extension for the investigation of the role of ligaments in the joint kinematic behavior was created and validated against the experimental data of 23 cadavers.

Special attention was paid to keep the numerical cost as low as possible. For both the tibial and patellar kinematics, the results computed with our model are inside the experimental corridor. Both the experimental and computational results obtained for the femoro-tibial and the femoro-patellar kinematics are also comparable to data reported in the literature. From a clinical perspective, the developed knee model constitutes a valuable tool for exploring new treatment strategies, such as the development of ACL substitutes for ligament reconstructions where it can be used to test different designs and evaluate the restoration of knee kinematics right after implantation.

References

1. D.M. Doroski, K.S. Brink, J.S. Temenoff, Techniques for biological characterization of tissue-engineered tendon and ligament. Biomaterials **28**(2), 187–202 (2007)
2. C. Azmy, S. Guérard, X. Bonnet, F. Gabrielli, W. Skalli, EOS® orthopaedic imaging system to study patellofemoral kinematics: assessment of uncertainty. Orthop. Traumatol. Surg. Res. **96**(1), 28–36 (2010)
3. G. Li, S. Zayontz, L.E. DeFrate, E. Most, J.F. Suggs, H.E. Rubash, Kinematics of the knee at high flexion angles: an in vitro investigation. J. Orthop. Res. **22**(1), 90–95 (2004)
4. A. Lorenz, O. Müller, P. Kohler, M. Wünschel, N. Wülker, U.G. Leichtle, The influence of asymmetric quadriceps loading on patellar tracking – an in vitro study. Knee **19**(6), 818–22 (2012)
5. R. Nagamine, L.A. Whiteside, T. Otani, S.E. White, D.S. McCarthy, Effect of medial displacement of the tibial tubercle on patellar position after rotational malposition of the femoral component in total knee arthroplasty. J. Arthroplasty **11**(1), 104–110 (1996)
6. M.S. Andersen, D.L. Benoit, M. Damsgaard, D.K. Ramsey, J. Rasmussen, Do kinematic models reduce the effects of soft tissue artefacts in skin marker-based motion analysis? An in vivo study of knee kinematics. J. Biomech. **43**(2), 268–273 (2010)
7. L. Lucchetti, A. Cappozzo, A. Cappello, U.D. Croce, Skin movement artefact assessment and compensation in the estimation of knee-joint kinematics. J. Biomech. **31**(11), 977–984 (1998)
8. C.B. Ma, K. Lee, M.A. Schrumpf, S. Majumdar, Analysis of three-dimensional in vivo knee kinematics using dynamic magnetic resonance imaging. Oper. Tech. Orthop. **15**(1), 57–63 (2005)
9. K. Suzuki, N. Hara, S. Mikami, T. Tomita, K. Iwamoto, T. Yamazaki et al., In vivo kinematic analysis of posterior-stabilized total knee arthroplasty for the valgus knee operated by the gap-balancing technique. Knee **21**(6), 1124–1128 (2014)
10. L. Blankevoort, R. Huiskes, A. De Lange, Recruitment of knee joint ligaments. J. Biomech. Eng. **113**(1), 94–103 (1991)
11. B.T. Feeley, M.S. Muller, A.A. Allen, C.C. Granchi, A.D. Pearle, Isometry of medial collateral ligament reconstruction. Knee Surg. Sports Traumatol. Arthrosc. **17**(9), 1078–1082 (2009)
12. M. Adouni, A. Shirazi-Adl, R. Shirazi, Computational biodynamics of human knee joint in gait: from muscle forces to cartilage stresses. J. Biomech. **45**(12), 2149–2156 (2012)
13. Z. Hao, D. Jin, Y. Zhang, J. Zhang, *A Finite Element 3D Model of In Vivo Human Knee Joint Based on MRI for the Tibiofemoral Joint Contact Analysis*, ed. by VG Duffy. Digital Human Modeling [Internet] (Springer, Berlin, 2007), pp. 616–22 [cited 2015 June 4]. http://link.springer.com/chapter/10.1007/978-3-540-73321-8_70
14. A. Kiapour, A.M. Kiapour, V. Kaul, C.E. Quatman, S.C. Wordeman, T.E. Hewett et al., Finite element model of the knee for investigation of injury mechanisms: development and validation. J. Biomech. Eng. **136**(1), 011002 (2013)

15. X. Liu, M. Zhang, Redistribution of knee stress using laterally wedged insole intervention: finite element analysis of knee–ankle–foot complex. Clin. Biomech. **28**(1), 61–67 (2013)
16. G. Papaioannou, G. Nianios, C. Mitrogiannis, D. Fyhrie, S. Tashman, K.H. Yang, Patient-specific knee joint finite element model validation with high-accuracy kinematics from biplane dynamic Roentgen stereogrammetric analysis. J. Biomech. **41**(12), 2633–2638 (2008)
17. C. Rao, C.K. Fitzpatrick, P.J. Rullkoetter, L.P. Maletsky, R.H. Kim, P.J. Laz, A statistical finite element model of the knee accounting for shape and alignment variability. Med. Eng. Phys. **35**(10), 1450–1456 (2013)
18. Y. Wang, Y. Fan, M. Zhang, Comparison of stress on knee cartilage during kneeling and standing using finite element models. Med. Eng. Phys. **36**(4), 439–447 (2014)
19. G. Limbert, M. Taylor, J. Middleton, Three-dimensional finite element modelling of the human ACL: simulation of passive knee flexion with a stressed and stress-free ACL. J. Biomech. **37**(11), 1723–1731 (2004)
20. N.A. Ramaniraka, P. Saunier, O. Siegrist, D.P. Pioletti, Biomechanical evaluation of intra-articular and extra-articular procedures in anterior cruciate ligament reconstruction: a finite element analysis. Clin. Biomech. **22**(3), 336–343 (2007)
21. F. Xie, L. Yang, L. Guo, Z. Wang, G. Dai, A study on construction three-dimensional nonlinear finite element model and stress distribution analysis of anterior cruciate ligament. J. Biomech. Eng. **131**(12), 121007 (2009)
22. E.M. Abdel-Rahman, M.S. Hefzy, Three-dimensional dynamic behaviour of the human knee joint under impact loading. Med. Eng. Phys. **20**(4), 276–290 (1998)
23. P. Beillas, G. Papaioannou, S. Tashman, K.H. Yang, A new method to investigate in vivo knee behavior using a finite element model of the lower limb. J. Biomech. **37**(7), 1019–1030 (2004)
24. M.A. Baldwin, C.W. Clary, C.K. Fitzpatrick, J.S. Deacy, L.P. Maletsky, P.J. Rullkoetter, Dynamic finite element knee simulation for evaluation of knee replacement mechanics. J. Biomech. **45**(3), 474–483 (2012)
25. W. Mesfar, A. Shirazi-Adl, Biomechanics of the knee joint in flexion under various quadriceps forces. Knee **12**(6), 424–434 (2005)
26. J. Noailles, *Modelisation Elements Finis De L'articulation Du Genou [Internet]* (ENSAM, Paris, 1999) [cited 2015 Jun 5]. http://www.theses.fr/1999ENAM0039
27. K. Choi, J.L. Kuhn, M.J. Ciarelli, S.A. Goldstein, The elastic moduli of human subchondral, trabecular, and cortical bone tissue and the size-dependency of cortical bone modulus. J. Biomech. **23**(11), 1103–1113 (1990)
28. R.Y. Hori, L.F. Mockros, Indentation tests of human articular cartilage. J. Biomech. **9**(4), 259–268 (1976)
29. A.A. Amis, A.M.J. Bull, C.M. Gupte, I. Hijazi, A. Race, J.R. Robinson, Biomechanics of the PCL and related structures: posterolateral, posteromedial and meniscofemoral ligaments. Knee Surg. Sports Traumatol. Arthrosc. **11**(5), 271–281 (2003)
30. D.L. Butler, M.D. Kay, D.C. Stouffer, Comparison of material properties in fascicle-bone units from human patellar tendon and knee ligaments. J. Biomech. **19**(6), 425–432 (1986)
31. N. Chandrashekar, H. Mansouri, J. Slauterbeck, J. Hashemi, Sex-based differences in the tensile properties of the human anterior cruciate ligament. J. Biomech. **39**(16), 2943–2950 (2006)
32. A. Race, A.A. Amis, The mechanical properties of the two bundles of the human posterior cruciate ligament. J. Biomech. **27**(1), 13–24 (1994)
33. J.R. Robinson, A.M. Bull, A.A. Amis, Structural properties of the medial collateral ligament complex of the human knee. J. Biomech. **38**(5), 1067–1074 (2005)
34. T. Sugita, A.A. Amis, Anatomic and biomechanical study of the lateral collateral and popliteofibular ligaments. Am. J. Sports Med. **29**(4), 466–472 (2001)
35. D.R. Wilson, J.D. Feikes, A.B. Zavatsky, J.J. O'Connor, The components of passive knee movement are coupled to flexion angle. J. Biomech. **33**(4), 465–473 (2000)

36. T. Zantop, T. Schumacher, N. Diermann, S. Schanz, M.J. Raschke, W. Petersen, Anterolateral rotational knee instability: role of posterolateral structures. Winner of the AGA-DonJoy Award 2006. Arch. Orthop. Trauma Surg. **127**(9), 743–752 (2007)
37. R. Philippot, J. Chouteau, R. Testa, B. Moyen, In vitro analysis of patellar kinematics: validation of an opto-electronic cinematic analysis protocol. Knee Surg. Sports Traumatol. Arthrosc. **18**(2), 161–166 (2010)

Chapter 3
Challenges to Validate Multi-Physics Model of Liver Tumor Radiofrequency Ablation from Pre-clinical Data

Chloé Audigier, Tommaso Mansi, Hervé Delingette, Saikiran Rapaka, Tiziano Passerini, Viorel Mihalef, Raoul Pop, Michele Diana, Luc Soler, Ali Kamen, Dorin Comaniciu, and Nicholas Ayache

3.1 Introduction

Subject-specific modeling of liver tumor radiofrequency ablation (RFA) can provide additional guidance to radiologists during the intervention and improve the planning of the procedure, which are challenged by inter-subject variability in tissue characteristics, the heterogeneous cooling effect of large neighboring vessels, porous circulation, and blood coagulation. More specifically, such models could help clinicians in deciding where to place the heating probe and for how long heating must be applied. Several approaches have been developed to describe and simulate RFA of liver tumors. They differ in their choice of the biophysical phenomena that are considered and the type of experimental data used to design and validate them. All simulations are based on the bioheat equation considering a cooling effect that is either diffuse [1] or localized at neighboring vessels [2–5]. Furthermore, the cooling effect due to venous flow in the liver parenchyma is also

C. Audigier (✉)
Inria Sophia-Antipolis, Asclepios Research Group, Sophia-Antipolis, France

Siemens Corporate Technology, Imaging and Computer Vision, Princeton, NJ, USA
e-mail: chloe.audigier@inria.fr

T. Mansi • S. Rapaka • T. Passerini • V. Mihalef • A. Kamen • D. Comaniciu
Siemens Corporate Technology, Imaging and Computer Vision, Princeton, NJ, USA

H. Delingette • N. Ayache
Inria Sophia-Antipolis, Asclepios Research Group, Sophia-Antipolis, France

R. Pop • M. Diana
IHU, Strasbourg, France

L. Soler
IHU, Strasbourg, France

IRCAD, Strasbourg, France

G. Joldes et al. (eds.), *Computational Biomechanics for Medicine*,
DOI 10.1007/978-3-319-28329-6_3

considered in [4, 5]. Few authors [2, 4, 5] have proposed to simulate RFA on realistic subject-specific geometries extracted from images and only [6] has personalized biophysical parameters on patient data in order to minimize the discrepancy between simulated and measured necrotic (ablated) regions.

Up to now, the comparison between simulated and measured necrotic regions has been used by several authors [2, 4–6] as the main criteria of success in predicting the effect of RFA on abdominal tumors, for either model validation or personalization. However, the necrosis of tissue is the resultant of several combined physical phenomena, mainly the heat transfer and cell death mechanisms, meaning that a given ablated region may be explained by several combinations of parameters. In addition to this identifiability issue, the size of the tumor extent can only be known reliably from post-operative imaging which makes it difficult to eventually update the ablation plan during the procedure. A method that relies also on pre-operative or interventional data for parameter identification is therefore required for RFA models to be clinically useful.

In this paper, an extended validation strategy of RFA is introduced, based also on delivered electrical power during ablation and temperature drop during cooling in addition to the extent of ablated regions. This approach leads to an increased confidence in the computed temperature map, a clinical surrogate for tissue damage during intervention. Furthermore, the probe temperature and delivered power are information that are readily available, in real-time, from the RF system and therefore could be used to update the therapy plan during the intervention.

The RFA computational model becomes subject-specific after three levels of personalization: anatomical, heat transfer, and cellular necrosis. The computational model, implemented using the Lattice-Boltzmann Method (LBM), relies on anatomies extracted from CT and device-based measurements (Sect. 3.3).

The proposed approach was successfully tested on seven surrogate hepatic tumors implanted on three healthy swine. Following the validation, we showed how this framework can be used for personalization. Parameter estimation was performed on two tumors to select a subject-specific set of parameters (Sect. 3.4). It leads to accurate predictions for both temperature evolution and necrotic region extension: mean error between measured and simulated temperature of 12.0 °C, mean point-to-mesh error between predicted and actual ablation extent of 3.5 mm.

3.2 Pre-clinical Study for Model Validation

3.2.1 Experimental Setup

Pigs are considered as a relevant animal model as their hepatic system is similar to the human one. The extensive pre-clinical study includes three swine. Several surrogate tumors (diameters < 3 cm) were inserted at various locations of the liver (close to vessels or Glisson capsule) under ultrasound (US) guidance, followed by the acquisition of pre-operative CT images including portal, venous, and arterial phases. The surrogate tumors were made of a specific gel which exhibited a

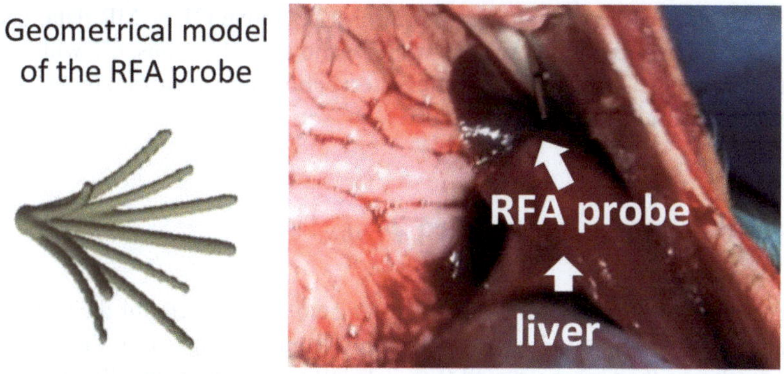

Fig. 3.1 (*Left*): mesh model of the probe with the nine tips derived from a CT image of the probe only; (*Right*): photo of the probe inside the pig liver

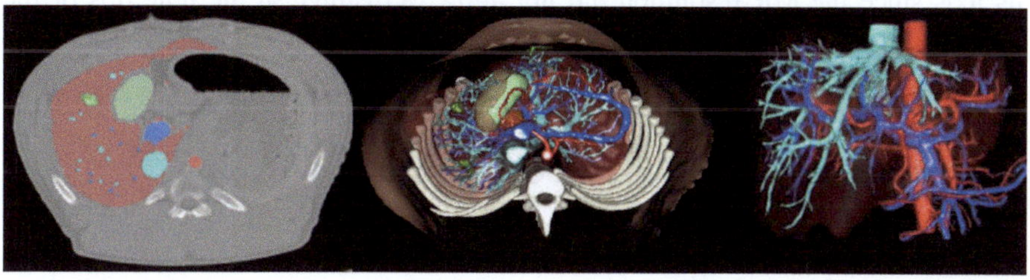

Fig. 3.2 Segmentation of the liver, arterial vessels, portal and hepatic networks, surrogate tumors, and biliary vesicle. (*Left*): overlay of the segmented areas on pre-operative CT image. (*Middle*): creation of the 3D model. (*Right*): vessels and liver parenchyma

hyper-intense signal in CT. An MR-compatible RFA probe (RITA, AngioDynamics) was deployed at 2 cm of diameter under US guidance (Fig. 3.1, right panel). An MR image was then acquired to get the position of the probe in the liver. The temperature and delivered power were monitored and recorded intra-operatively during and after the ablation. Finally, a post-operative CT or T2 MRI was acquired 2 days after the ablation to assess the extent of the necrotic areas. Overall, pre-, intra-, and post-operative images were available, along with interventional device measurements. To the best of our knowledge, no such validation setup has been reported in the literature.

3.2.2 Data Preprocessing

From the pre-operative CT data, the following anatomical structures were segmented semi-automatically by experts (Visible Patient, Strasbourg, France) and meshed (Fig. 3.2): the parenchyma, all tumors, hepatic veins, vena cava, portal vein, biliary vesicle, and hepatic arteries. From post-operative CT or MR data, necrotic areas were segmented and meshed as well. Due to ethical reasons, CT with contrast agent could not be acquired 2 days after intervention, making accurate pre- to post-operative registration nearly impossible.

3.2.3 3D Modeling of the RFA Probe

The geometry of the probe when deployed at 2 cm (the diameter of the area defined by the tips of the probe is 2 cm) was acquired from a CT image of the probe only. A 3D mesh was then reconstructed (Fig. 3.1, left panel) and manually registered on the pre-operative CT using the main axis of the probe and the intra-operative MR data. The rigid registration was visually checked by an expert up to a rotation along its axis since the MR resolution did not allow to distinguish the nine tips of the probe.

3.3 Subject-Specific RFA Model

3.3.1 Heat Transfer Model

The bioheat equation describes how the heat flows from the probe through the liver while taking into account the cooling effect of the main vessels, as proposed in the Pennes model [7]. The temperature T is computed by solving the following reaction–diffusion equation:

$$\rho_t c_t \frac{\partial T}{\partial t} = Q + \nabla \cdot (d_t \nabla T) + R(T_{b0} - T) \qquad (3.1)$$

where ρ_t, c_t, d_t are the density, heat capacity, conductivity of the tissue, Q, the source term, R, the reaction coefficient, and T_{b0}, the blood temperature (assumed constant) in large vessels. In this study, the blood flow within the parenchyma considered as a porous medium is not taken into account as how it affects the overall cooling has not been clearly quantified in the literature, in particular when compared to the diffusive effect. Nonetheless our model includes the cooling effect, also called heat sink effect, of all hepatic large vessels (veins and arteries). Equation (3.1) is solved using the LBM for fast computation on general purpose Graphics Processing Units (GPU). LBM has been developed for CFD and is now a well-established discretization method. Verification of the implementation has been performed through a comparison with an analytical solution and it has the advantage over FEM to be easily parallelized in GPU [4]. An isotropic Cartesian grid with 7-connectivity topology and Neumann boundary conditions at the boundaries of the liver is employed. A Multiple-Relaxation-Time (MRT) model is used for increased stability [8]. For a time step of $\Delta t = 0.5$ s, faster than real-time computation can be achieved on a standard desktop machine (Windows 7, Intel Xeon, 3.30 GHz, 16 GB RAM, 12 CPUs, Nvidia Quadro K5000 4.0 GB).

3.3.2 Cellular Necrosis Model

A three-state model [9] is coupled with the bioheat equation to compute tissue necrosis. Each cell has a probability to be either undamaged (U), vulnerable (V), or necrotic (N). Those probabilities vary with the simulated temperature over time according to the following state equation:

$$U \underset{\gamma}{\overset{\beta(T)}{\rightleftarrows}} V \xrightarrow{\delta(T)} N \tag{3.2}$$

where $\beta(T) = \bar{\beta}e^{T/T_k}$, $\delta(T) = \bar{\delta}e^{T/T_k}(1 + 10N)$, and γ are the transition rates. Unlike in [10], three distinct transition rates are considered to allow cells to reach the vulnerable state. Figure 3.3 represents the solution at one vertex over time if a constant temperature of 105 °C is applied. The cell death model is strongly coupled to the bioheat equation. More specifically, it depends on the temperature computed based on the bioheat model (the transition rates $\beta(T)$ and $\delta(T)$ are temperature dependent), whose parameters depend on the state of the cell (the heat capacity depends on the state of the cell: c_t^U, c_t^V, and c_t^N correspond, respectively, to the heat capacity of undamaged, vulnerable, and necrotic tissue).

Equation (3.2) gives three coupled ODEs that are solved with a first order explicit scheme on the same grid and with the same time step as the bioheat equation. For the heat transfer and cell death models, parameters are initially set to values from the literature [7], Table 3.1 reports them. The conductivity d_t depends on the temperature through $d_t = \bar{d}_t * (1 + 0.00161 * (T - 310))$ as in [5].

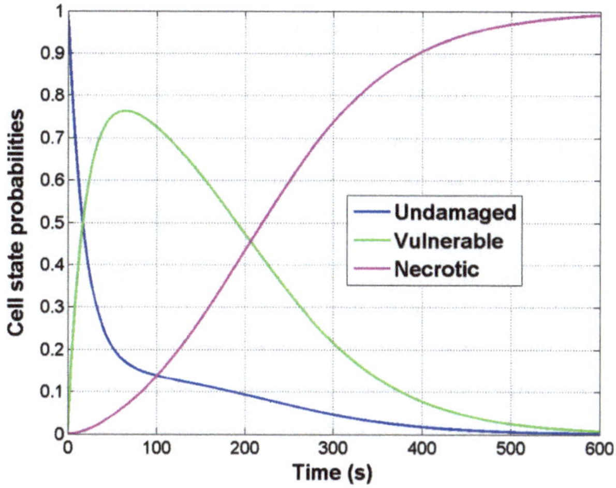

Fig. 3.3 Cell state evolution over time when tissue is heated at 105 °C

Table 3.1 Nominal value of model parameters

Notation	Parameter name	Nominal value
γ (s^{-1})	Recovery rate coefficient	7.7×10^{-3}
$\bar{\beta}$ (s^{-1})	Damage rate coefficient	3.3×10^{-3}
$\bar{\delta}$ (s^{-1})	Vulnerable rate coefficient	3.3×10^{-3}
c_t^V (J(kg K)$^{-1}$)	Heat capacity of vulnerable tissue	3.6×10^{3}
c_t^N (J(kg K)$^{-1}$)	Heat capacity of necrotic tissue	0.67×10^{3}
c_t^U (J(kg K)$^{-1}$)	Heat capacity of undamaged tissue	3.6×10^{3}
$\bar{d}_t$ (W(m K)$^{-1}$)	Heat conductivity	0.512

3.3.3 *Parameter Estimation from Probe Measurements*

Computation of Heating Power During the intervention, the delivered power and the temperature distribution are measured by the ablation probe itself. We assume that the measured power is actually strongly correlated (proportional) to the heat power $P(t)$ delivered through radiofrequency to heat the liver tissue. Proportionality is assumed to account for power dissipation due to electrical resistance, and the unknown surface ratio of the probe being heated. The heat power $P(t)$ delivered to the tissue can be computed at each time step of the simulation from the bioheat equation according to Fourier's law:

$$P(t) = \int_S d_t \frac{\partial T(t)}{\partial \boldsymbol{n}} dS \tag{3.3}$$

where S is the probe surface and $\boldsymbol{n}$ is the outer normal at that surface. During our experiments, the measured electrical power appeared to reach its maximum during the increase in temperature, before reaching the plateau of 105 °C (Fig. 3.6). These phenomena cannot be explained by a constant heat capacity which would lead to a peak after the plateau is reached. Instead, this observation suggests that the cells reach their vulnerable state faster. Thus we update our necrosis model accordingly: after studying the ODEs of the model and a sensitivity analysis on its parameters, $\bar{\delta}$ has been modified such that tissues reach very fast their vulnerable state which entails a significant change of heat capacity (Table 3.3).

Cooling Stage RFA is simulated by imposing the measured temperature at the tips of the probe (Dirichlet boundary condition) until the heating stops at a given time $t = t_a$. In the absence of any delivered heat power, the nine tips of the probe cool down at a speed which depends on the conductivity d_t and the heat capacity c_t. Thus during the cooling period $t_c - t_a$ (cooling stops at time $t = t_c$), the simulated temperature T_s can be compared with the measurements T_m read from five tips of the probe (four tips do not have any thermistors).

Personalization After a sensitivity analysis, we chose to estimate the heat capacity c_t^U and the constant part of the conductivity $\bar{d}_t$ as they mainly influence the delivered power, the temperature distribution, and the size of ablated regions. These two parameters are personalized from probe measurements (temperature and power) by minimizing the following cost function:

$$f = \frac{1}{t_a} \sum_{t=0}^{t=t_a-1} \frac{(P_m(t) - P_s(t))^2}{\sigma_{P_m}^2} + \frac{1}{t_c - t_a} \sum_{t=t_a}^{t=t_a+t_c} \frac{(m_{T_m}(t) - m_{T_s}(t))^2}{\sigma_{T_m}^2} \tag{3.4}$$

where σ_{Pm} and σ_{Tm} are the standard deviations associated with the heat power and the temperature, both of them evaluated from the variability in the available observations (equal to 13.3 and 5 in our experiments). To cope with the uncertainty in the rotation of the probe along its axis, the mean tip temperature m_T is used for the personalization.

Using DAKOTA,[1] a gradient-free optimization method, the Constrained Optimization BY Linear Approximations (COBYLA), is used to minimize the cost function Eq. (3.4) as only a few forward simulations (typically 20) are required. COBYLA is a sequential trust-region algorithm. Initially, the total domain of parameters is visited (here, d_t: 0.51 to 6.14, c_t^U: 18 to 3618) and then the region is contracted. The estimation took around 13 h (25 iterations).

3.4 Evaluation on Swine Data

3.4.1 Model Validation

The model was evaluated on seven ablations performed in three swine. The applied RFA protocol was not exactly the same for all ablations. Five ablations were performed through several short cooling and heating periods, whereas the other two ablations included only one long final cooling stage after a continuous heating period. For all pigs, nominal value of parameters (reported in Table 3.1) was employed. In each case, the simulated lesion was compared with the registered ground-truth. Due to the uncertainty in the registration of the post-operative image to the pre-operative image, the necrotic lesion segmented on the post-operative image is registered rigidly to the pre-operative image by aligning its barycenter with the barycenter of the simulated necrosis. Figure 3.4 shows results for tumor 2-1, in this case, the model under-estimates the lesion size. Quantitatively, point-to-mesh errors [11] computed between the simulated lesion and the registered post-operative ground-truth showed good prediction of the necrotic extent (3.5 mm of mean point-to-mesh error). Values are reported in Table 3.2. The prediction of the necrosis extent was valid up to 5 mm which can be considered as sufficient for clinical applications. Qualitatively, as one can see in Fig. 3.5, the simulated heat power and temperature were close to the heat power and the temperature given by the RFA probe itself.

[1] http://dakota.sandia.gov—multilevel framework for sensitivity analysis.

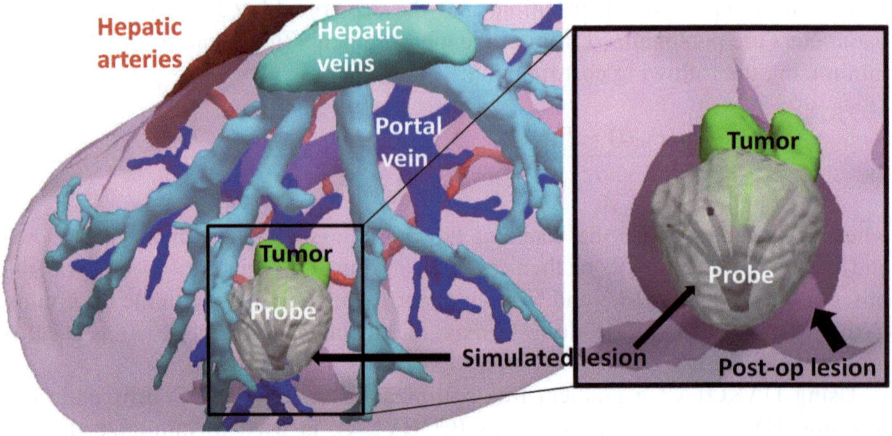

Fig. 3.4 Comparison between the simulated and the post-operative necrotic areas on tumor 2-1. (*Left*): the simulated lesion is showed around the RFA probe in the subject-specific geometry. (*Right*): zoom on the ablation area, the simulated lesion is qualitatively close to the registered post-operative lesion

Table 3.2 Evaluation on pig data

Pig	Necrotic point-to-mesh error (mm)
1	3.71 ± 2.49
2-1	4.06 ± 1.59
2-2	4.02 ± 1.35
2-3	2.46 ± 1.35
3-1	3.80 ± 2.39
3-2	4.85 ± 2.13
3-3	2.13 ± 1.71

3.4.2 Towards Model Personalization

As the data came from healthy pigs of similar age and weight, we hypothesized that the parameters would be the same for all of them. The personalization based on the probe temperature and power was performed for the two tumors with final cooling stage as it was long enough to observe reliably the effect of the conductivity d_t, yielding two sets of personalized heat capacity and conductivity values, reported in Table 3.3. We validated not only the ability to personalize the model but also its predictive power by evaluating the simulation results on five different tumors. Briefly, the minimization of the error between measured and simulated values of power and temperature was done only on two tumors (two different pigs). In both cases, the values independently found were really close: the same value of heat capacity was estimated, and the conductivity values were almost equal to the nominal value, as expected as the pigs were healthy.

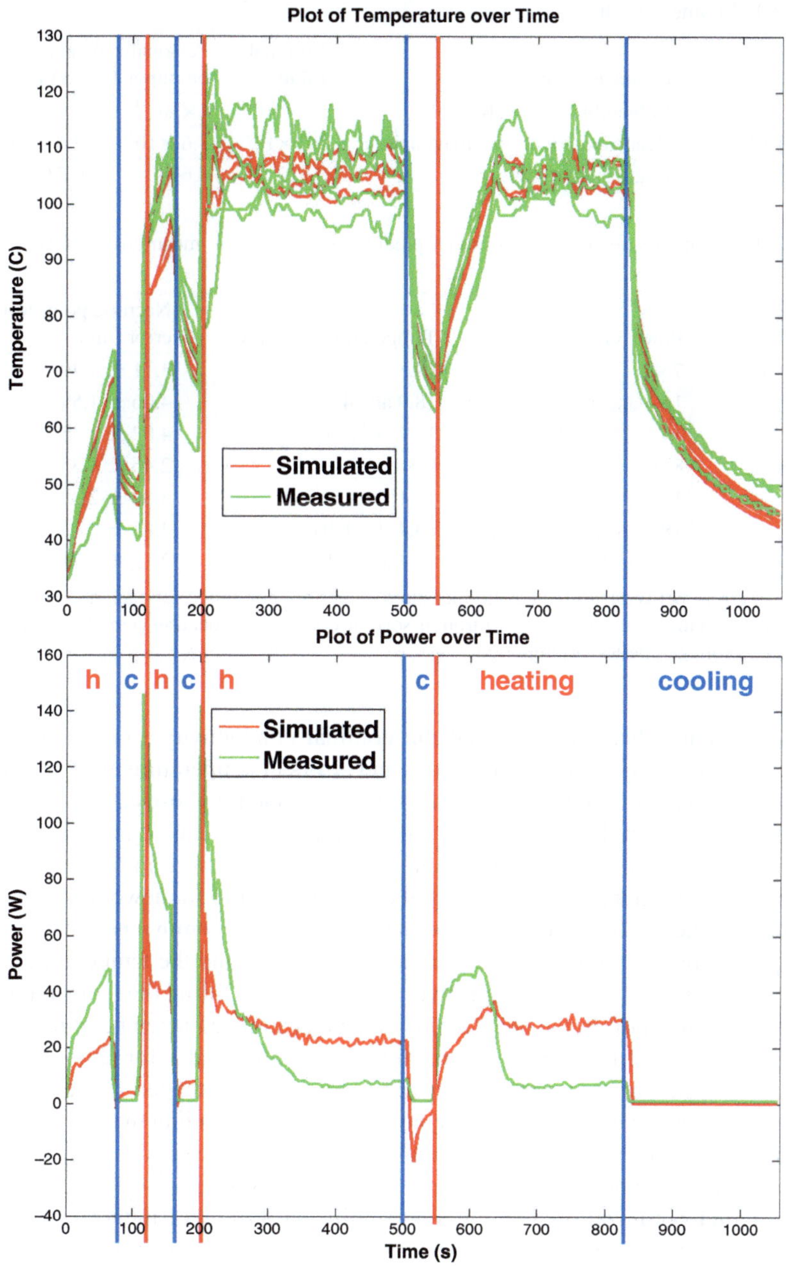

Fig. 3.5 Comparison between the simulated and the measured temperature on tumor 3-1. During the heating phase, the temperature is imposed, the power is simulated and compared to the measured one. During the cooling phase, the simulated (non-imposed) temperature is compared to the measured one

Table 3.3 Parameters values

Notation	Parameter name	Nominal value	Personalized values on tumor 1	on tumor 2-3
$\bar{\delta}$ (s^{-1})	Vulnerable rate coefficient	3.3×10^{-3}	1×10^{-4}	
c_t^U (J(kg K)$^{-1}$)	Heat capacity of undamaged tissue	3.6×10^{3}	3.6×10^{1}	3.6×10^{1}
$\bar{d}_t$ (W(m K)$^{-1}$)	Heat conductivity	0.512	0.614	0.512

Table 3.4 Quantitative evaluation of model prediction: on average, mean errors of 14.8 W and 12.0 °C

Pig	Power error (W)	Temperature error (°C)	Necrotic point-to-mesh error (mm)
1 (perso)	7.9	19.3	3.71 ± 2.49
2-1	16.5 and 19.2	18.9 and 4.7	4.06 ± 1.59
2-2	14.1, 15.5, 12.7	11.4, 7.3, 11.4	4.02 ± 1.35
2-3 (perso)	8.9	0.5	2.46 ± 1.35
3-1	10.3, 31.8, 20.2, 16.8	19.1, 21.9, 6.8, 3.7	3.80 ± 2.39
3-2	18.4, 16.2, 6.4	23.4, 13.0, 9.6	4.85 ± 2.13
3-3	8.5 and 14.4	19.4 and 1.1	2.13 ± 1.71

The mean errors between simulated and measured values are computed for each ablation. For the five ablations which were performed through several cooling and heating periods, the different error values for each period are reported

The estimated heat capacity and the nominal conductivity were then used to simulate RFA on the five remaining cases and errors in temperature, heat power, and necrosis size were evaluated. Small errors were obtained in those cases too, without previously having fit the parameters for those tumors. As detailed in Sect. 3.3.3, the vulnerable rate coefficient was also adjusted to match the raise in delivered power. These results confirmed the stability of the personalization framework. As reported in Table 3.4, the use of personalized parameters instead of nominal parameters led to good predictions of the necrosis extent, the heat power, and the temperature (mean errors of 14.8 W and 12.0 °C, respectively). However, by fitting the temperature and the delivered heat power, the novel approach could estimate the temperatures around each tumor at any time during the ablation. This additional information could be used as surrogate to assess the amount and location of damaged tissue (cells that received excessive heat but without being necrosed) surrounding the ablated region.

The extent of the necrotic area is mainly controlled by the conductivity d_t and the heat capacity of vulnerable cells c_t^V (not of undamaged ones c_t^U). Since d_t was not changed after optimization, there was no significant difference in terms of necrotic area after personalization, despite a better match for the measured power and cooling temperature as illustrated in Fig. 3.6. It was actually not surprising that the nominal d_t value was optimal since all pigs were healthy, of similar age and weight. Moreover the point-to-mesh errors were less than 5 mm; it suggests that simulations with optimized d_t and c_t^U are realistic in terms of necrotic area, power, and temperature

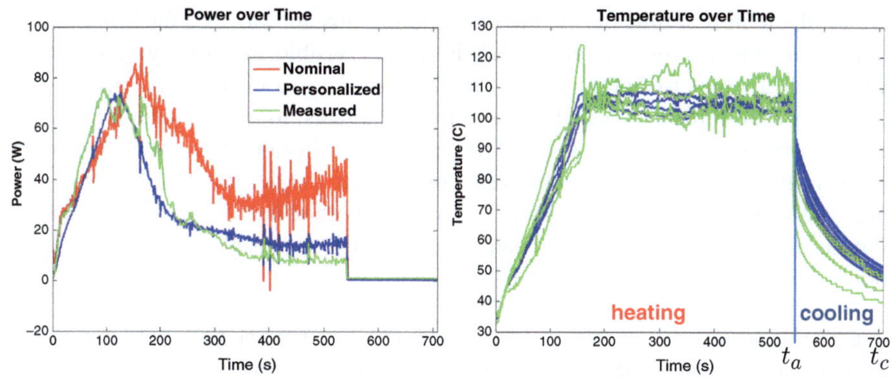

Fig. 3.6 Personalization of the heat capacity and the conductivity using (*Left*) the delivered power curve over ablation and (*Right*) the temperature distribution after ablation from tumor 1 (here, the *red* and *blue* curves are superimposed). The error between the measurement and the computation is reduced from 22.9 to 7.9 W for the power and is equal to 19.3 °C for the temperature

predictions which was the objective. In clinical settings, due to the large variety of diseases treated by RFA (cirrhosis, fibrosis, etc.), the proposed method should be suitable to get subject-specific parameters from easily accessible data.

3.5 Conclusion

In this paper, we proposed a novel approach to validate computational models of RFA based on pre-, intra-, and post-operative images and device-based measurements, in close-to-clinical settings. The approach was successfully evaluated on three swine and seven ablations. We presented a pre-clinical validation of a detailed model, required prior to any clinical study for which personalization would occur during the intervention.

In clinical RFA of liver tumors, we cannot assume that the biophysical parameters are the same for all patients, as assumed in this experiment for the three pigs, and personalization is therefore required. However, by evaluating the discrepancy in terms of temperature and delivered power, key biophysical parameters (the heat capacity and the conductivity of the bioheat model) could be estimated leading to promising predictions. Here, we just relied on information localized at the tips of the probe, the cooling temperature and the delivered power, during the intervention (no temperature maps, no post-op images required) and we showed that the model can be personalized from patient data, which was not granted. We limited the number of personalized parameters (2). However by estimating more parameters like c_t^V, c_t^N, current errors like the lesion size around tumor 2-1, for example, could be reduced.

This opens new perspectives for updating intra-operatively the RFA model prediction of each lesion based on those two probe measurements. Because they are available in real-time during the intervention, this appears as a far better option than using the necrosis extent which is mostly visible post-operatively. Future work will attempt to optimize the computation time to estimate on the fly key parameters from readily available probe measurements, to include additional cases and observations quantities (MR thermometry, for instance) in order to further improve the personalization and to reliably estimate the deformation between pre- and post-operative imaging for a precise validation of the necrosis extent prediction. Further refinements in the model can be added such as the advection effect of the porous circulation in the parenchyma [6].

Acknowledgements Part of this work was funded by the IHU in Strasbourg through the SimulAB project: Radio-frequency ablation on Liver. The authors are grateful to Gael Fourré, Franck Blindauer, Mourad Bouhadjar, and Rodrigo Cararo, radiology technicians and veterinary at the IHU Surgical Institute, for their valuable assistance in performing the experimental procedures.

References

1. X. Chen, G.M. Saidel, Mathematical modeling of thermal ablation in tissue surrounding a large vessel. J. Biomech. **131**, 011001 (2009)
2. I. Altrogge, T. Preusser, T. Kroger, S. Haase, T. Patz, R.M. Kirby, Sensitivity analysis for the optimization of radiofrequency ablation in the presence of material parameter uncertainty. Int. J. Uncertain. Quantif. **2**, 295–321 (2012)
3. Y. Jiang, S. Mulier, W. Chong, M. Diel Rambo, F. Chen, G. Marchal, Y. Ni, Formulation of 3D finite elements for hepatic RFA. Int. J. Model. Identif. Control. **9**, 225–235 (2010)
4. C. Audigier, T. Mansi, H.C Delingette, S. Rapaka, V. Mihalef, E. Boctor, M. Choti, A. Kamen, N. Ayache, D. Comaniciu, Efficient Lattice Boltzmann solver for patient-specific radiofrequency ablation of hepatic tumors. IEEE Trans. Med. Imaging **34**(7), 1576–1589 (2015). doi:10.1109/TMI.2015.2406575
5. S. Payne, R. Flanagan, M. Pollari, T. Alhonnoro, C. Bost, D. O'Neill, T. Peng, P. Stiegler, Image-based multi-scale modelling and validation of radio-frequency ablation in liver tumours. Philos. Trans. R. Soc. A **369**, 4233–4254 (2011)
6. C. Audigier, T. Mansi, H. Delingette, S. Rapaka, V. Mihalef, D. Carnegie, E. Boctor, M. Choti, A. Kamen, D. Comaniciu et al., Parameter estimation for personalization of liver tumor radiofrequency ablation, in *MICCAI Workshop on Abdominal Imaging–Computational and Clinical Applications*, 2014
7. H.H. Pennes, Analysis of tissue and arterial blood temperatures in the resting human forearm. J. Appl. Physiol. **85**, 5–34 (1998)
8. C. Pan, L.S. Luo, C.T. Miller, An evaluation of lattice boltzmann schemes for porous medium flow simulation. Comput. Fluids **35**, 898–909 (2006)
9. M. Breen, X. Chen, D. Wilson, G. Saidel, Modeling cellular thermal damage from radiofrequency ablation, in *EMBS/BMES Conference* (IEEE, New York, 2002)
10. D. O'Neill, T. Peng, P. Stiegler, U. Mayrhauser, S. Koestenbauer, K. Tscheliessnigg, S. Payne, A three-state mathematical model of hyperthermic cell death. Ann. Biomed. Eng. **39**, 570–579 (2011)
11. Y. Zheng, A. Barbu, B. Georgescu, M. Scheuering, D. Comaniciu, Fast automatic heart chamber segmentation from 3D CT data using marginal space learning and steerable features, in *IEEE 11th International Conference on Computer Vision, ICCV 2007* (IEEE, New York, 2007), pp. 1–8

Chapter 4
Robust Landmark Identification for Generating Subject Specific Models for Biomechanics

Duane T.K. Malcolm, Habib Y. Baluwala, Poul M.F. Nielsen, and Martyn P. Nash

4.1 Introduction

Tracking breast tumours between different imaging modalities and positions within the breast can provide important clinical information. Tracking can be aided by using predictive biomechanics models that compute the deformation of the breast between different states. Typically, finite element methods (FEM) are used to simulate deformations of the breast [1], however generating the mesh for the simulation is a manual and laborious process. As medical imaging and computational biomechanics become more common, so will the need for robust automated methods to generate meshes in a reasonable amount of time. This study describes a robust automated method for identifying landmarks in a magnetic resonance (MR) scan. The position of landmarks can guide image segmentation [2] and the placement of nodes for mesh generation.

Landmarks can be defined as distinct anatomical points in the body or as edges that can be used as feature points [3]. Landmark detection in medical imaging is focused on three approaches: (a) machine learning algorithms [4, 5]; (b) image intensity models [6]; and (c) invariant geometric measures such as curvatures and extreme points based on their shapes [7]. Hartkens et al. [8] investigate 3D differential operators for the detection of point landmarks in 3D MR and CT images. Frantz et al. [9] implemented a statistical approach for landmark detection that requires the user to set an initial region of interest in the vicinity of the

D.T.K. Malcolm • H.Y. Baluwala
Auckland Bioengineering Institute, University of Auckland, Auckland, New Zealand

P.M.F. Nielsen • M.P. Nash (✉)
Auckland Bioengineering Institute, University of Auckland, Auckland, New Zealand

Department of Engineering Science, University of Auckland, Auckland, New Zealand
e-mail: martyn.nash@auckland.ac.nz

G. Joldes et al. (eds.), *Computational Biomechanics for Medicine*,
DOI 10.1007/978-3-319-28329-6_4

landmark. Most of these methods are either semi-automatic or cannot accurately detect landmarks on the skin surface from torso MR images due to the wide dynamic intensity range or variability of torso shapes across patients and populations. Furthermore, some of these methods are sensitive to noise and result in large errors in identifying the position of landmarks.

Baluwala et al. [10] used statistical shape models [11] to improve the reliability of template matching techniques for identifying landmarks. This study extends this method to three-dimensions and improves the reliability by using a partial least-squares regression (PLSR) model to predict the positions and the template images of the landmarks. This reduces the search region, which reduces computational time; provides better initial conditions for the search method; and increases the likelihood of correlation with the actual landmark.

In this paper, we describe the data set used to train the statistical models, the training methods, the landmark search methods, and the application and analysis of the predictive ability of this method. This proposed method is compared with the standard template matching method in order to contrast the accuracy and robustness of the proposed method. The tracheal bifurcation and jugular notch at the top of the sternum are used as test landmarks.

4.2 Methods

4.2.1 Overview of the Data

The landmark detection process was trained and tested on 51 MR scans (Fig. 4.1a) acquired with the subject lying prone (face-down) in a clinical MR scanner (Siemens Skyra T1-weighted, pixel size: 0.84–1.07 mm, and slice thickness: 0.9–1.1 mm). The subjects were all females with the metrics shown in Table 4.1.

The proposed method was tested using two landmarks, the tracheal bifurcation and the jugular notch. The landmarks were manually identified in each of the 51 MR images. This is the only manual process which is required for the training process.

4.2.2 Training the Process

The training process involved the generation of statistical models of the relative landmark positions and landmark template images, four partial least-square regression (PLSR) models to predict the centre of shape model, the relative landmark positions, and the weights for the statistical template image for each landmark. The statistical models of the shape models are used to predict the initial search region and search for the landmark positions in the template matching process. The PLSR

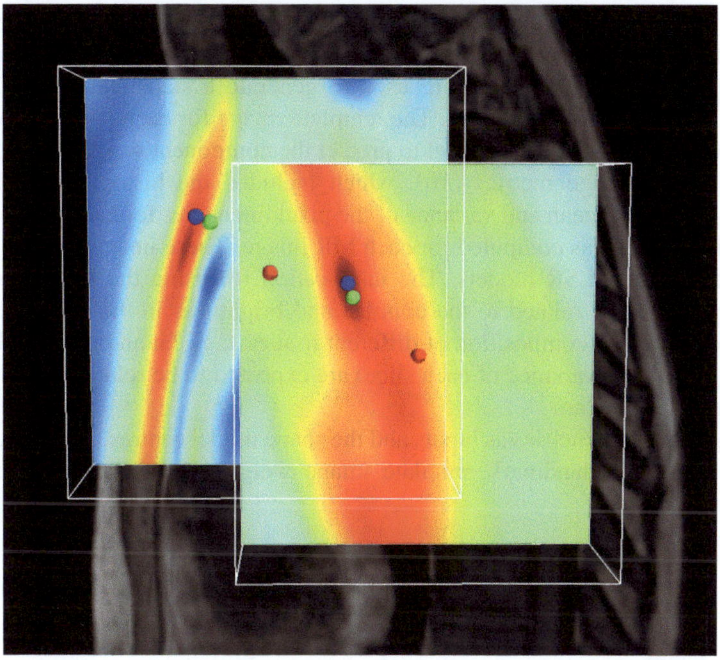

Fig. 4.1 Sagittal MR image of the female torso. Respectively, the *red*, *blue*, and *green spheres* indicate the initial prediction, final prediction, and actual positions of the tracheal bifurcation (*right*) and jugular notch (*left*). Superimposed are the overlapping correlation fields from cross-correlating the predicted landmark template images with the search regions. The correlation fields are normalised between 0 and 1

Table 4.1 Statistics relating to the subjects used in this study

Metric	Mean ± SD	Range
Age	40 ± 15 yo	19–69 yo
Height	1.65 ± 0.07 m	1.48–1.78 m
Weight	67 ± 10 kg	42–89 kg

models of the template image weights are used to predict the template images based on the image metrics. These statistical models are described in more detail below.

For thetracheal bifurcation (TB) and jugular notch (JN) landmark set, the relative landmark locations (X_S) and the centre of the landmarks (x_c) were computed by

$$X_S = [x_{TB} - x_c, x_{JN} - x_c]$$

$$x_c = (x_{TB} + x_{JN})/2$$

A statistical shape model of the relative landmark positions (X_S) was generated with three components using principal components analysis (PCA).

Additionally, a 48 pixels cubed (1 mm^3 voxels) template image (I_T) was extracted for each landmark where the image was centred on the landmark position. The population of template images are used to generate a statistical template image with two components using PCA. The template image for each landmark was then projected back into the PCA model to predict the component weights (w_{TB}, w_{JN}).

Finally, the image metrics (M_I), which included the histogram of the pixel intensities and the mean and variance of the pixel intensities for each axial, coronal, and sagittal slice, was computed for each MR image. These image metrics are used as inputs for the PLSR models. The justification for using the histogram is that it is expected to be related to the contrast and brightness in the scan, the size of the subject, and the composition of different tissues, e.g., fat and muscle. The slice metrics (mean and variance of each slice) are expected to be related to the position of the body in the scan.

Using the image metrics as inputs and the shape model centres, relative landmark positions, and the landmark template image weights, four PLSR models were generated:

$$M_I \rightarrow \mathrm{PLSR_C} \rightarrow x_c \tag{4.1}$$

$$M_I \rightarrow \mathrm{PLSR_S} \rightarrow X_S \tag{4.2}$$

$$M_I \rightarrow \mathrm{PLSR_{TB}} \rightarrow w_{TB} \tag{4.3}$$

$$M_I \rightarrow \mathrm{PLSR_{JN}} \rightarrow w_{JN} \tag{4.4}$$

4.2.3 *Identification of Unseen Landmarks*

Given a new MRI scan (the "unseen" image), there are three steps to identify the locations of the tracheal bifurcation and jugular notch: (1) make initial predictions of the shape model centre, relative landmark positions, and template image for each landmark; (2) cross-correlate the template image with the search region; and (3) search of the correlation fields for the maximum combined correlation where the sampled locations are constrained by the statistical shape model. Each step is described in more detail below.

4.2.3.1 Initial Predictions

First, the image metrics (M_I) of the unseen MR image were computed and fed into the PLSR models described in Sect. 4.2.2 to predict the shape model centre (x_c), the relative landmark positions (X_S), and weights (w_{TB}, w_{JN}) for the statistical template models of the tracheal bifurcation and jugular notch.

The predicted landmark positions (X_L) in the MR scan were then given by

$$X_L = X_S + x_c$$

The landmark template images (I_T) for the tracheal bifurcation and jugular notch were reconstructed from the associated statistical template images using the weights (w_{TB}, w_{JN}) predicted above.

4.2.3.2 Cross-Correlation

Cross-correlation was performed using the template matching methods described in OpenCV library [12] but extended to three-dimensions. The cross-correlation method was used with normalisation and the mean of the image subtracted. The search region was centred on each of the predicted landmark positions (X_L) where the region was $160 \times 160 \times 160$ mm (1 mm^3 voxels). The predicted landmark template images were cross-correlated with their respective search regions to obtain correlation fields of each landmark (see Fig. 4.1).

4.2.3.3 Statistical Shape Model Search

The final predictions of the landmark positions were obtained by performing a search where the objective function was to maximise the product of the correlation values samples at the positions reconstructed from the statistical shape model and where the degrees-of-freedom were the shape model centre (x_c) and statistical shape model weights (w_s).

Prior to the search, a better prediction of the centre of the shape model (x_c) was obtained by multiplying the correlation fields and searching for the position of the maximum correlation product (see Fig. 4.2). The initial statistical shape model weights (w_S) were predicted by projecting the predicted relative landmark positions (X_S) into the statistical shape model. The "fmin_l_bfgs_b" function in the scipy.optimize module [13] was used to minimise the negative of the product of the correlation values sampled at the landmark positions. The minimisation function is a multivariate bound constrained method that uses the L-BFGS-B algorithm [14]. This function is used in order to constrain the weights of the shape model to within ± 3 standard deviations.

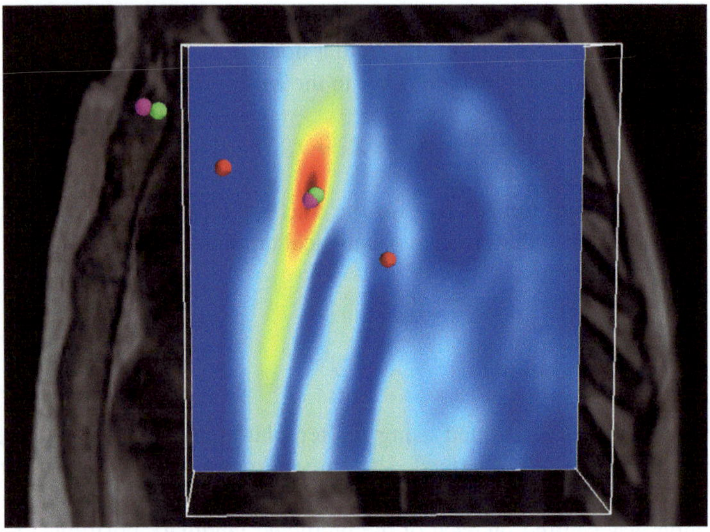

Fig. 4.2 The product of the normalised correlation fields for the tracheal bifurcation and jugular notch landmarks (shown in Fig. 4.1). The *red spheres* are the initial landmark positions predicted using PLSR. These are adjusted to the *magenta spheres* based on the maximum of the combined correlation fields. The *green spheres* are the actual landmark positions identified in the image

4.2.4 Testing the Algorithm

The predictive ability of the proposed method was tested using a leave-one-out approach. Each subject in turn was selected as the *unseen* subject, and the remaining subjects were used to train algorithm. This trained algorithm was then used to predict the positions of the landmarks in the unseen MR image. The error for each landmark prediction was calculated as the Euclidean distance between predicted and manually identified landmark positions. The leave-one-out approach was applied in turn to each of the 51 MRI scans. The histogram of the errors and the percentage of errors over 10 mm for the leave-one-out cases provided an indication of the predictive ability (accuracy and robustness) of the landmark identification algorithm. The 10 mm failure rate threshold was chosen near the tail of the distribution of errors (see Figs. 4.3 and 4.4) in order to categorise the cases that failed to find the landmark. This threshold was simply used to indicate the robustness of the method, rather than the accuracy of the method.

The proposed method was compared to the template matching method for each individual landmark where the position of the landmark was taken as the position of the maximum correlation.

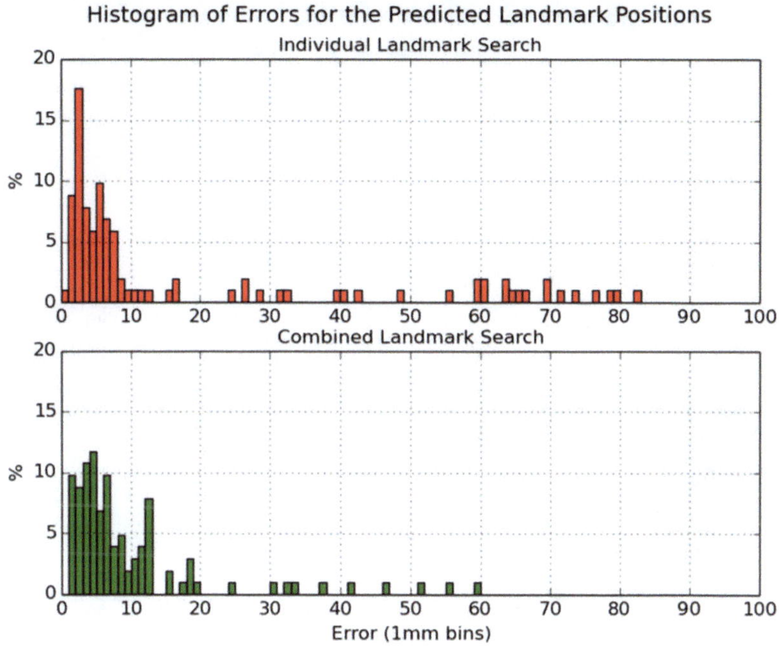

Fig. 4.3 The distribution of errors for the prediction of the tracheal bifurcation and jugular notch when using the individual (*top*) and combined (*bottom*) landmark search methods. The template landmark images and initial positions and shape model was based on the average across the population ($n = 51$)

4.3 Results and Discussion

4.3.1 Identifying the Landmark Positions

Two landmark search methods were compared:

1. Individual landmark search, which uses template matching of individual landmarks where the landmark positions are identified by the locations of maximum correlation from the template matching.
2. Combined landmark search, which uses template matching and statistical models where the landmark positions are identified by the locations of maximum combined correlation from the template matching using the fitted statistical model.

For each method, the template landmark image and initial landmark positions were taken as the average from the population in the first case, and are predicted using PLSR in the second case. Each search took between 15 and 20 s on an Intel Core i7-2620M 2.7 GHz CPU.

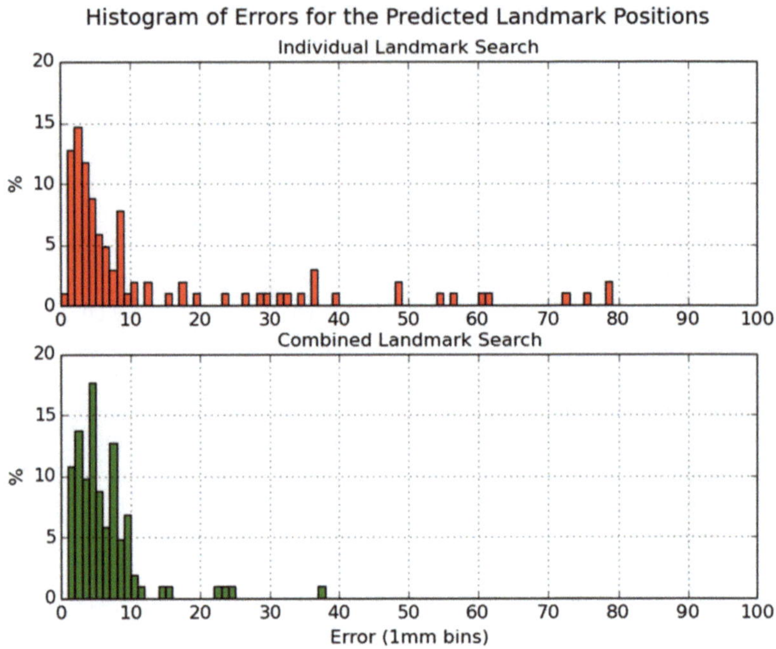

Fig. 4.4 The distribution of errors in the prediction of the tracheal bifurcation and jugular notch when using the individual (*top*) and combined (*bottom*) landmark search methods. The template landmark images and initial positions and shape model was based on the average across the population

Table 4.2 The error and failure rate (error > 10 mm) using average and predicted initial conditions and template image for individual and combined landmark search methods

Error (mm)	Average		Predicted using PLSR	
Individual	18.9 ± 24.7	33 % > 10 mm	13.9 ± 19.1	28 % > 10 mm
Combined	10.2 ± 11.7	31 % > 10 mm	6.2 ± 5.2	9 % > 10 mm

The results are summarised in Table 4.2, which shows the mean error and percentage of errors above 10 mm for each search method with and without predicted initial landmark positions and template image. The worst result was given by the template matching method for individual landmark using the average landmark locations and mean template image. The best result was given by the combined landmark search method with predicted initial landmark positions and predicted template image. The improvements in the error and the percentage over 10 mm were over threefold.

Figures 4.3 and 4.4 provide a better indication of the distribution of the errors and failures for the average and predicted initial conditions, respectively. The histograms show progressive improvements given by each method, and with or without predicted landmark positions and template images.

We investigated the influence of two factors - predicting the initial landmark positions, and predicting the template landmark image - on the accuracy and robustness of the combined search method. Predicting the template image, but not the initial landmark positions, gave a predicted landmark error of 8.5 ± 10.0 mm where 21.6 % of the errors were greater than 10 mm. Predicting the initial landmark positions, but not the template image, gave a predicted landmark error of 7.7 ± 8.5 mm where 20.6 % of the errors were greater than 10 mm. Individually, the use of predictions gave similar improvements in accuracy and robustness, however predicting both the initial landmark positions and the template landmark image gave a significantly better cumulative result.

4.3.2 Future Work

The results from this study are promising, but the proposed method involves many variables that can be tuned to improve the accuracy and robustness of the predictions. These factors include the number and choice of landmarks, the resolution, voxel size and processing (gradient and smoothing) of template images, the types of metrics used in the PLSR models, the number of modes used in the statistical shape models, and the type metric used in the template matching (cross-correlation, Mahalanobis distance, squared-errors, or a combination). Given the improvement seen related to predicting the initial landmark positions and template images, it would be worth investigating other metrics for the PLSR model, such as Haar features [15], and other prediction techniques.

Given that the proposed method provides robust and accurate identification of the tracheal bifurcation and jugular notch, these landmarks can be used to perform a subsequent search for additional landmarks. This work is in progress and is providing robust and accurate identification of the sternal angle, intervertebral joints, the location of the armpit, the centre line of the ribs, and the skin surface (see Fig. 4.5). The intention is to use these landmarks to generate subject specific finite element meshes for biomechanical simulations.

The proposed method was developed and tested on 51 images with well controlled imaging parameters, but it was less robust with MR images that exhibited significant imaging artefacts. However, with future development we are optimistic that the robustness of this method can be improved to cope with the imaging artefacts and variations, and possibly, various imaging modalities.

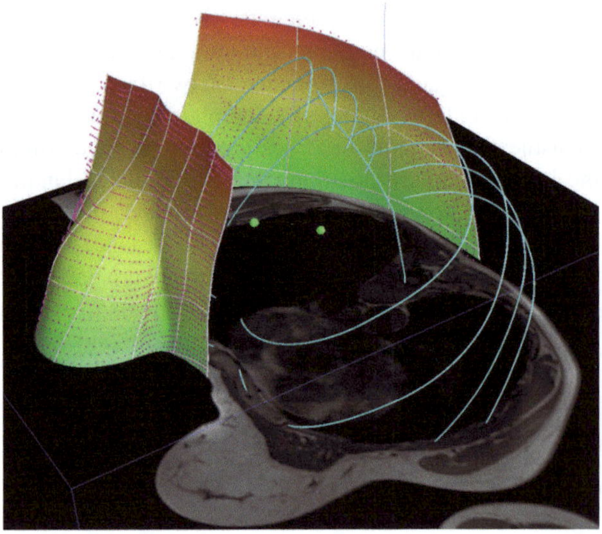

Fig. 4.5 Work in progress to extend this method to identify other landmarks (sternal angle and intervertebral joints), the *centre lines* of the ribs, and the skin surface with a fitted mesh

4.4 Conclusions

The presented landmark identification method that combines template matching and statistical models, and uses PLSR to predict the initial landmark positions and template landmark images, was able to predict the landmark positions with an error of 6.2 ± 5.2 mm with 9.1 % of the errors over 10 mm. This is a significant improvement over template matching alone, which gave a prediction error of 18.9 ± 24.7 mm with 33 % of the predictions over 10 mm. It was found that predicting both the initial landmark positions and template landmark images significantly improved the robustness of the method. Work in progress to extend this method to other landmarks including the spine, ribs, and skin surface has given promising initial results. It is expected with further development and the use of other feature detection and learning techniques, the accuracy and robustness of this method can be improved.

References

1. T.P. Babarenda Gamage, V. Rajagopal, P.M. Nielsen, M.P. Nash, Patient-specific modelling of breast biomechanics with applications to breast cancer detection and treatment, in *Patient-Specific Modelling in Tomorrow's Medicine* (Springer, Berlin, 2012), pp. 379–412
2. Y. Zheng, A. Barbu, B. Georgescu, M. Scheuering, D. Comaniciu, Four-chamber heart modeling and automatic segmentation for 3-D cardiac CT volumes using marginal space learning and steerable features. IEEE Trans. Med. Imaging **27**(11), 1668–1681 (2008)
3. A. Mostayed, R.R. Garlapati, G.R. Joldes, A. Wittek, A. Roy, R. Kikinis, K. Miller, Biomechanical model as a registration tool for image-guided neurosurgery: evaluation against B-spline registration. Ann. Biomed. Eng. **41**(11), 2409–2425 (2013)

4. A. Criminsi, J. Shotton, S. Bucciarelli, Decision forests with long range spatial context for organ localisation in CT volumes, in *Medical Image Computing and Computer-Assisted Intervention (MICCAI)* (Springer, Berlin, 2009), pp. 69–80
5. V. Potesil, T. Kadir, G. Platsch, M. Brady, Improved anatomical landmark localization in medical images using dense matching of graphical models, in *Proceedings of the British Machine Vision Conference (BMVC)*, vol. 4, no. 6 (British Machine Vision Association, 2010), p. 9
6. S. Worz, K. Rohr, Localization of anatomical point landmarks in 3D medical images by fitting 3D parametric intensity models. Med. Image Anal. **10**(1), 41–58 (2006)
7. K. Rohr, Extraction of 3D anatomical point landmarks based on invariance principles. Pattern Recogn. **32**, 3–15 (1999)
8. T. Hartkens, K. Rohr, H.S. Stiehl, Evaluation of 3D operators for the detection of anatomical point landmarks in MR and CT images. Comput. Vis. Image Underst. **86**(2), 118–136 (2002)
9. S. Frantz, K. Rohr, H.S. Stiehl, Development and validation of a multi-step approach to improved detection of 3D point landmarks in tomographic images. Image Vis. Comput. **23**(11), 956–971 (2005)
10. H.Y. Baluwala, D.T.K. Malcolm, J.W.Y. Jor, P.M.F. Nielsen, M.P. Nash, Automatic landmark detection using statistical shape modelling and template matching, Computational Biomechanics for Medicine (Springer International Publishing, 2014), pp. 75–82
11. T.F. Cootes, C.J. Taylor, Statistical models of appearance for medical image analysis and computer vision, in *Medical Imaging* (International Society for Optics and Photonics, 2001), pp. 236–248
12. G. Bradski, OpenCV, Dr. Dobb's Journal of Software Tools (2000)
13. F. Pedregosa et al., Scikit-learn: machine learning in Python. J. Mach. Learn. Res. **12**, 2825–2830 (2011)
14. R.H. Byrd, P. Lu, J. Nocedal, A limited memory algorithm for bound constrained optimization. SIAM J. Sci. Stat. Comput. **16**(5), 1190–1208 (1995)
15. P. Viola, M. Jones, Rapid object detection using a boosted cascade of simple features, in *Proceedings of the 2001 IEEE Computer Society Conference on Computer Vision and Pattern Recognition*, 2001

[illegible]

Chapter 5
Forward Problem of Time-Resolved Diffuse Optical Tomography Considering Biological Tissue Deformation

A.Y. Potlov, T.I. Avsievich, S.V. Frolov, and S.G. Proskurin

5.1 Introduction

Near infrared optical irradiation which is usually used in diffuse optical tomography (DOT) is diffusely transmitted through a biological tissue carries useful information about the shape, size, location, and optical properties of the tissue internal structure, i.e. gives possibility to map tissue optical properties [1, 2]. However, to use this information it is necessary to solve forward and inverse problems of irradiation propagation in such media.

Because of complex distribution of scattering and absorption properties, that have different shape, size, location, etc., in general case the inverse problem doesn't have the exact solution [1–4]. Therefore, it is necessary to study the regularities of light propagation through simulated objects, that describe the simplified standard cases that correspond to location of tumors, hematomas, hygromas, and other inhomogeneities. On this basis, we have developed and optimize approximate radiative transfer models, according to which, absorption or scattering are the key factors of light attenuation which is passing through a biological tissue.

The importance of such regularities and approximate radiative transfer models is that they allow to understand better the impact of various factors on propagation and attenuation of optical radiation in biological tissues. It is known that the external shape of the tissue can significantly affect the quality of image reconstruction in DOT [4]. Therefore, it is necessary to provide a good, tight contact of optical fibers to inject and detect diffusely transmitted photons in the investigated area. It can significantly reduce the loss of the useful signal, and thus to improve the accuracy of absorption and scattering properties mapping. However, in some cases, this also leads to a change of the shape (deformation) of the tissue. In this regard,

A.Y. Potlov (✉) • T.I. Avsievich • S.V. Frolov • S.G. Proskurin
Tambov State Technical University, Tambov, Russia
e-mail: zerner@yandex.ru

G. Joldes et al. (eds.), *Computational Biomechanics for Medicine*,
DOI 10.1007/978-3-319-28329-6_5

compression plates usually appear as the cause of the deformation [5]. In our case it is an elastic bracelet to fixture the irradiation source and detector fibers. Application of these plates or the bracelet leads to the image artifacts, caused by the discrepancy between the actual boundaries of investigated area and its mathematical model representation [6–8]. Because of susceptible to deformation, these artifacts can cause a significant problem in the study of a soft tissue such as breast [9, 10].

Considering spatial distribution of scattering and absorption coefficients, authors of paper [6] describe influence of the female breast deformation to the results of DOT inverse problem solution. To reduce the number of artifacts in the reconstructed image, in the iterative process of forward problem solution it is suggested to consider change of the object's shape [6]. For this purpose, the shape of the breast is determined using 3D camera. Then, using computer simulation, deformations due to the pressure of the source and detector fibers are predicted. Mathematical model for the simulation is based on the basic equation of the theory of elasticity.

Interesting approach of mammography-based elastography is suggested in the paper [8] for breast tumors diagnostics. The key feature of the method is the female breast elastogram simulation on the basis of the traditional X-ray mammography or DOT. The special attention in this method is paid to the criterions which help to distinguish malignant tumors from the healthy tissue by their elastic properties [8].

Paper [11] describes computer methods to make simulation of elastic properties of the breast tissue for the purpose of surgical biopsy. Suggested model allows to determine tumor position in the deformed object using MRT data for the undeformed object. This approach is based on the small deformation theory, considering that big area deformations are to be divided into small ones.

The purpose of this work is to analyze influence of biological tissue deformations on the optical properties, propagation and attenuation, of infrared irradiation in it.

5.2 The Model of Diffuse Migration of Photons

To describe the diffusion of photons in biological tissues we used the Model of a Drop—the calculation and visualization of photon density normalized maximum (PDNM) motion in a slice of a biomedical object. After being injected, a single ultra short pulse of irradiation with predetermined number of photons diffusely moves inside an object like a drop of ink in the water [12–14].

This model allows to describe the experimental data for homogeneous and inhomogeneous cases, and to visualize PDNM. It is based on the numerical solution of radiative transfer equation (RTE) in the diffusion approximation for a light pulse with a fixed number of photons.The diffusion approximation for RTE describes the balance of energy in a medium containing scattering particles [4] and expressed in partial differential equations of parabolic type:

$$\begin{aligned} &\frac{1}{c}\frac{\partial \phi\left(x,y,z,t\right)}{\partial t} - D\left(x,y,z\right)\nabla^{2}\phi\left(x,y,z,t\right) \\ &+ \mu_{a}\left(x,y,z\right)\phi\left(x,y,z,t\right) = S\left(x,y,z,t\right), \forall x,y,z \in \Omega, \end{aligned} \tag{5.1}$$

where $c = \frac{c_0}{\nu_{\text{object}}}$—speed of light in the medium; c_0—speed of light in vacuum; ν_{object}—the relative refractive index of the simulated object (Ω) and its boundaries ($\partial\Omega$); x, y, z—coordinates of all points of the final simulated area, consisting of an inner part of the simulated object Ω, its boundaries $\partial\Omega$, radiation source (q), detectors, and the medium, that is surrounding the object; $D(x, y, z) = \{3[\mu_a(x, y, z) + (1 - g(x, y, z))\mu_s(x, y, z)]\}^{-1}$ and $\mu_a(x, y, z)$—diffusion and absorption coefficients, respectively; $\mu_s(x, y, z)$—the scattering coefficient; point of interest is determined by the coordinates—x, y, z; g—anisotropy factor (the average cosine of the scattering angle); $\phi(x, y, z, t)$—the photon density at the point with coordinates x, y, z at a time t; and $S(x, y, z, t)$—photon source function.

Robin boundary condition is used for a description of the photon flux at the boundary Ω of the simulated object [15–17]:

$$\phi(x, y, z, t) + 2D(x, y, z)F\frac{\partial\phi(x, y, z, t)}{\partial\widehat{n}(x, y, z)} = 0, \forall x, y, z \in \partial\Omega, x, y, z \notin q, \qquad (5.2)$$

here $\widehat{n}(x, y, z)$—direction of the outer normal to the boundary $\partial\Omega$ at the point with coordinates x, y, z. F—Fresnel reflection coefficient [3, 15], calculated as:

$$F = \frac{\frac{2}{1-R_0} - 1 + |\cos(Q_c)|^3}{1 - |\cos(Q_c)|^2},$$

where R_0 and Q_c coefficients, respectively, equal to:

$$R_0 = \frac{\left(\frac{\nu_{\text{object}}}{\nu_{\text{medium}}} - 1\right)^2}{\left(\frac{\nu_{\text{object}}}{\nu_{\text{medium}}} + 1\right)^2} \text{ and } Q_c = \arcsin\left(\frac{\nu_{\text{medium}}}{\nu_{\text{object}}}\right),$$

where ν_{medium}—the relative refractive index of the medium surrounding the object (for the air $\nu_{\text{medium}} = 1$).

After completion of the simulation iterative process PDNM function, $\phi(x, y, z, t)$, is normalized with respect to its maximum $\phi_{\max}(x, y, z, t)$:

$$\phi_{\text{norm}}(x, y, z, t) = \frac{\phi(x, y, z, t)}{\phi_{\max}(x, y, z, t)},$$

and represented as follows:

$$\phi_{\text{PDNM}}(x, y, z, t) = \left| \begin{array}{l} 1, \phi_{\text{norm}}(x, y, z, t) \geq P \\ \phi_{\text{norm}}(x, y, z, t), \text{else} \end{array} \right.,$$

where P is the experimentally determined minimum of photon density level $0 < P \leq 1$ [18].

5.3 Biomechanical Properties Modeling of Soft Biological Tissues

In terms of biomechanics, soft biological tissues are nonlinear elastic media. Nevertheless, in some cases, for example, when the deforming forces cause small bending ($\leq$ 5%), the soft tissues can still be considered as a media with linear properties [6–8]. For example, in DOT, the breast tissue should be presented as a linear isotropic pseudo-incompressible medium.

In this case, the basic equation of the elasticity theory for quasi-static deformation on the internal nodes of the simulated area is given by [6, 8]:

$$(\lambda + \mu) \nabla (\nabla u) + \mu \nabla^2 u = 0, \tag{5.3}$$

and the same equation for nodes on the boundary $\partial\Omega$ of the studied area is represented as follows [6]:

$$\left((\lambda + \mu) \nabla (\nabla u) + \mu \nabla^2 u\right) \cdot \breve{n} = h, \tag{5.4}$$

where $\breve{n}$—a unit vector directed outwards from Ω; h—represents the tension on the surface and boundary of the simulated area; $u = (u_1, u_2, u_3)$—the displacement vector components at the axes x, y, z in the Cartesian coordinate system; and μ and λ—Lame's elastic constants. These constants for isotropic medium (first and second Lame's elastic constants) are associated with the Young's modulus E and Poisson ratio ν as follows [8–10]:

$$\mu = \frac{E}{2(1+\nu)}$$

and

$$\lambda = \frac{\nu E}{(1+\nu)(1-2\nu)}.$$

Thus, the simulated object is considered to be free of any initial deformations and an influence of the internal deforming forces (such as muscle activity). All deformations are considered to be caused by external loads, such as source and detection fibers and their fixtures.

Note that, scattering and absorption coefficients of the tissue under the deformation should change but insignificantly [6, 8]. The suggested model does not consider these changes.

To describe the mathematical model it is necessary to give consideration to the experimental setup for time-resolved DOT (Fig. 5.1) as well as bracelet technicalities which could cause female breast deformations in the points of source and detector fibers attachment:

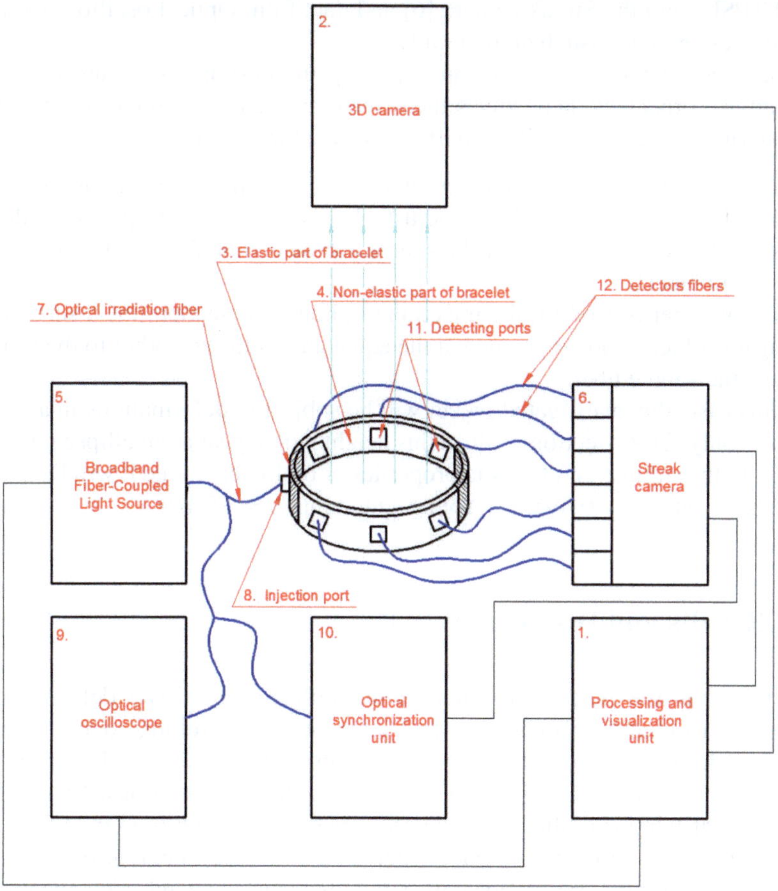

Fig. 5.1 Block diagram of the experimental setup for time-resolved diffuse optical tomography

1. By the signal of Processing and visualization unit [1] 3D camera [2] reads and determines the surface of the investigated object;
2. Weak reflecting and partially elastic bracelet is attached to the investigated object [3, 4];
3. Broadband light source [5] and Streak camera [6] are activated by the signal from Processing and visualization unit [1];
4. Broadband light source [5] generates a light pulse which goes to the Injection port [8] through the fibers. Part of the optical irradiation goes to the Optical oscilloscope [9] and Optical synchronization unit [10];
5. Diffusely transmitted through the object irradiation goes to the Detecting ports [11] then goes to the Detectors fibers [12] and to the detecting array of the Streak camera [6]. The latter registers the full set of time-resolved data—Time Point Spread Functions (TPSF);

6. All TPSF from the Streak camera [6] and data from Optical oscilloscope [9] go to Processing and visualization unit [1];
7. Processing and visualization unit [1] using the designed software makes calculations, processes data, solves the inverse problem, and maps absorption and scattering properties distribution hidden inside the object.

The important feature of the experimental setup is that the fibers are located on slightly reflective elastic band with adjustable diameter consisting of two identical halves. It allows fixing painlessly the source and detection fibers on the investigated biomedical object.

Only one fiber is used for the injection of photons, detection fibers surround the investigated object, and are located at the equal angles to each other to the right and left from the source fiber.

In this case, the investigated object will be subject to deformations in a plane of the fibers only. Cross-section made at this height will appear as an ellipse rather than a circle. In the simulation of elastic properties of biological tissue it will be assumed that Poisson ratio $\nu = 0.495$ and Young's modulus $E = 20\ kPa$.

5.4 Results and Discussion

Our study has the purpose to analyze the influence of tissue deformation on the optical irradiation propagation and attenuation. Therefore, to minimize 3D numerical simulation interpolation distortions, the FDM grid should be uniform; i.e. distance between the nodes should be the same in all three directions. To realize this the numerical solution of the Eqs. (5.1)–(5.4) was performed using Finite Difference Method, using the implicit difference scheme built on seven-point grid pattern [14].

The results of the simulations in a homogeneous and inhomogeneous non-deformed conical object are shown in Fig. 5.2. Distribution of optical properties of the slice of the object is shown in Fig. 5.3. It is taken at the half height of the object and in the plane of the fibers.

Several series of computer simulations have been performed to determine influence of deformation to the character of PDNM movement. Photon density distribution was consequently simulated ($P = 0.995 - 0.999$) for the time-resolved cases using the pulsed irradiation in homogeneous and inhomogeneous, deformed and undeformed conical objects with optical properties of breast tissue.

It has been found that, in all homogeneous cases PDNM moves to the geometric center of the conical object, regardless of the presence or absence of deformations, and also the values $\mu_a(x, y, z)$ and $\mu_s(x, y, z)$ [18].

The simulation results of the photon distribution in the plane (made at the level of the source and detector fibers) of homogeneous undeformed object at different moments of time t are shown in Fig. 5.4.

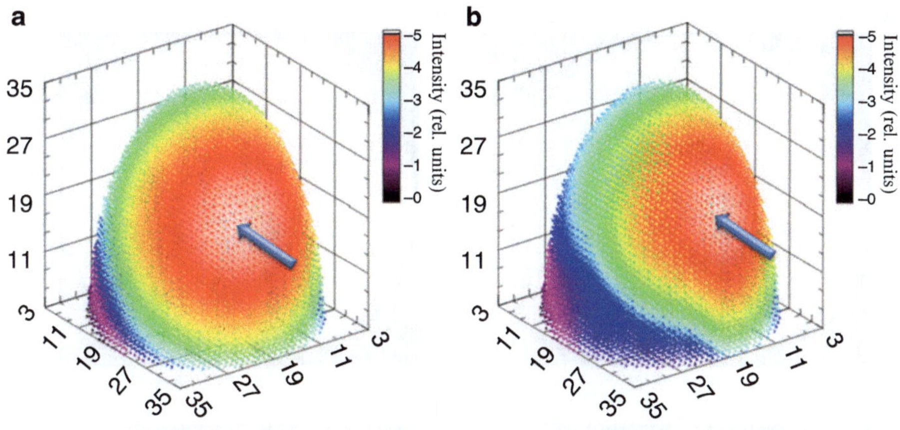

Fig. 5.2 The distribution of photons in homogeneous (**a**) and inhomogeneous (**b**) conical objects in 0.75 ns after the light pulse injection. Dimensions of the pictures are 136 mm × 136 mm × 136 mm

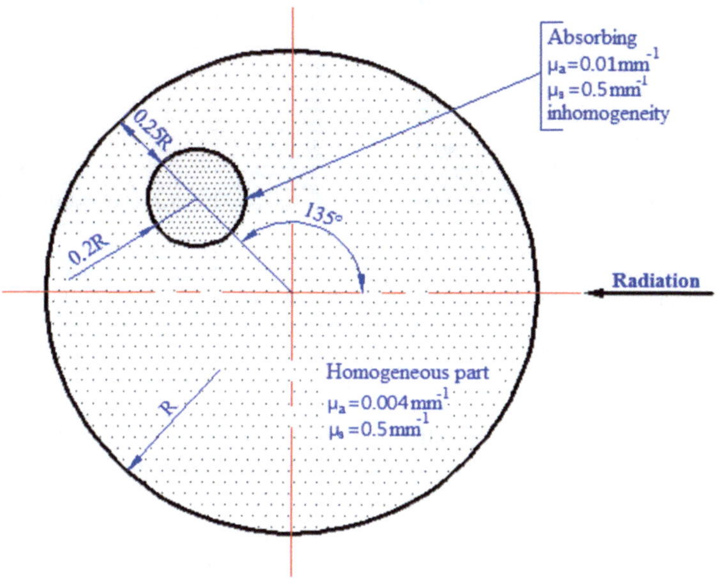

Fig. 5.3 Optical properties distribution in the plane taken at the half height of heterogeneous conical object. Inhomogeneity is considered to be spherical

Absorption, $\mu_a(x, y, z)$, and reduced scattering, $\mu_s^{'}(x, y, z)$, coefficients for $\forall r \in \Omega$ are equal to 0.004 and 0.5 mm^{-1}, respectively. Similar results for the optically homogeneous deformed object are shown in Fig. 5.5.

In the case when object has an absorbing inhomogeneity, the character of PDNM movement considerably varies in comparison with the homogeneous cases.

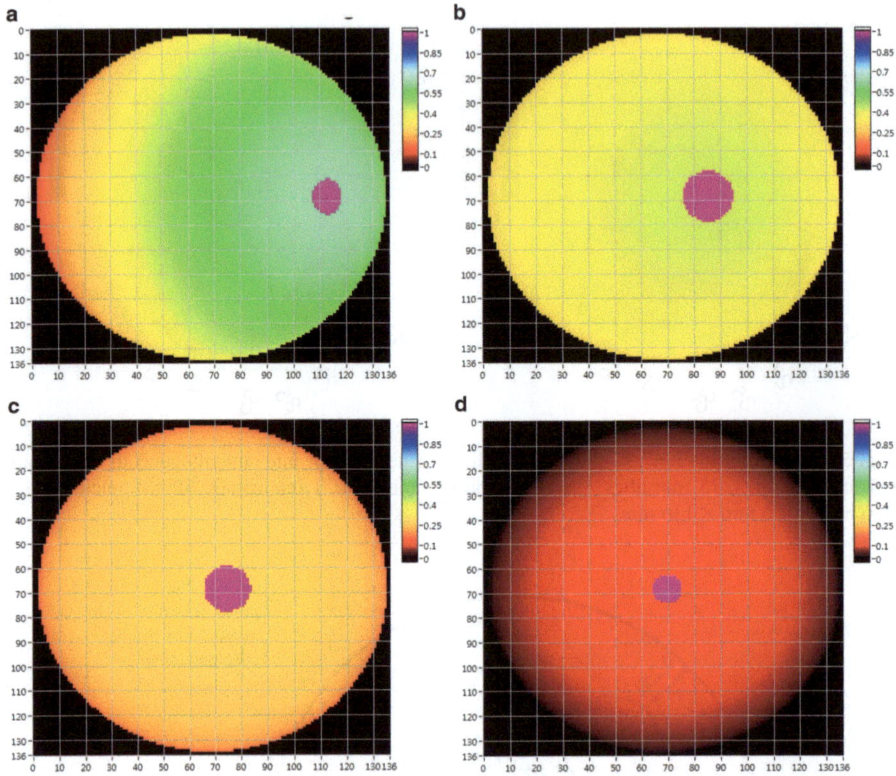

Fig. 5.4 Photon density distributions in the slice of the homogeneous conical object at the following times after the injection of the pulse: 0.7 ns (**a**), 2.1 ns (**b**), 3.5 ns (**c**), and 4.9 ns (**d**). *Purple spot* represents photon density normalized maximum in all cases. Dimensions of the pictures are 102 mm × 102 mm

Figure 5.6 shows photon density distributions in the plane at the half height of the inhomogeneous undeformed conical object at different time points. The size of the object is the same as in the homogeneous case [18].

Absorbing inhomogeneity with $\mu_a(x, y, z) = 0.01$ mm^{-1} is located at the angle of 135° with respect to the axis of the incident irradiation, at the depth of 0.25 of the plane radius, R. It represents a sphere with the diameter of 0.4 of the radius.

Figure 5.6 shows that in the case of absorbing inhomogeneity, PDNM moves toward the point, which is symmetrical to geometrical center of the heterogeneity relative to the center of the investigated object.

As a result of additional simulations, it was found that the speed of PDNM movement increases with the increasing of the inhomogeneity absorption coefficient and its size [19].

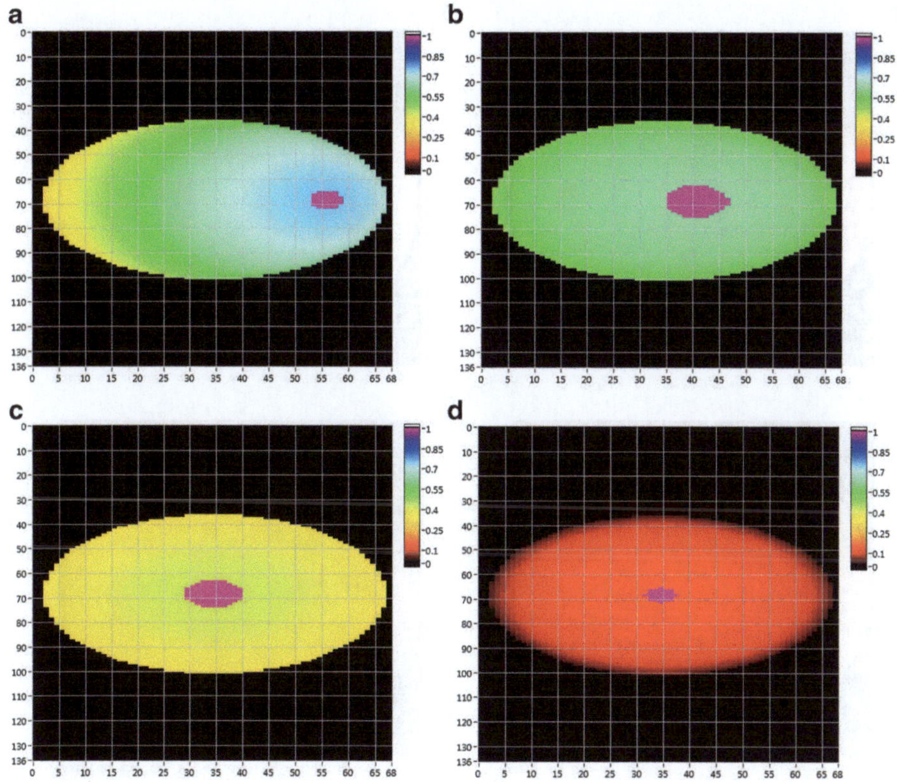

Fig. 5.5 Photon density distributions in the slice of the deformed conical homogeneous object at the following times after the injection of the pulse: 0.7 ns (**a**), 2.1 ns (**b**), 3.5 ns (**c**), and 4.9 ns (**d**). PDNM moves to the center. Dimensions of the pictures are 51 mm× 102 mm

Similar results were obtained for the deformed object, see Fig. 5.7. The figure shows that PDNM moves similarly, i.e. deformation of the object and the imbedded inhomogeneity, has no significant influence on the photon density and its normalized maximum.

However, rate of the radiation intensity decay in the cases with undeformed objects is slightly higher (8–10 %) than that in the cases with deformations. That means the deformations influence significantly on the overall distribution of the photon density except for the location of its maximum. Therefore they should be taken into account when solving the DOT inverse problem [6–8, 20].

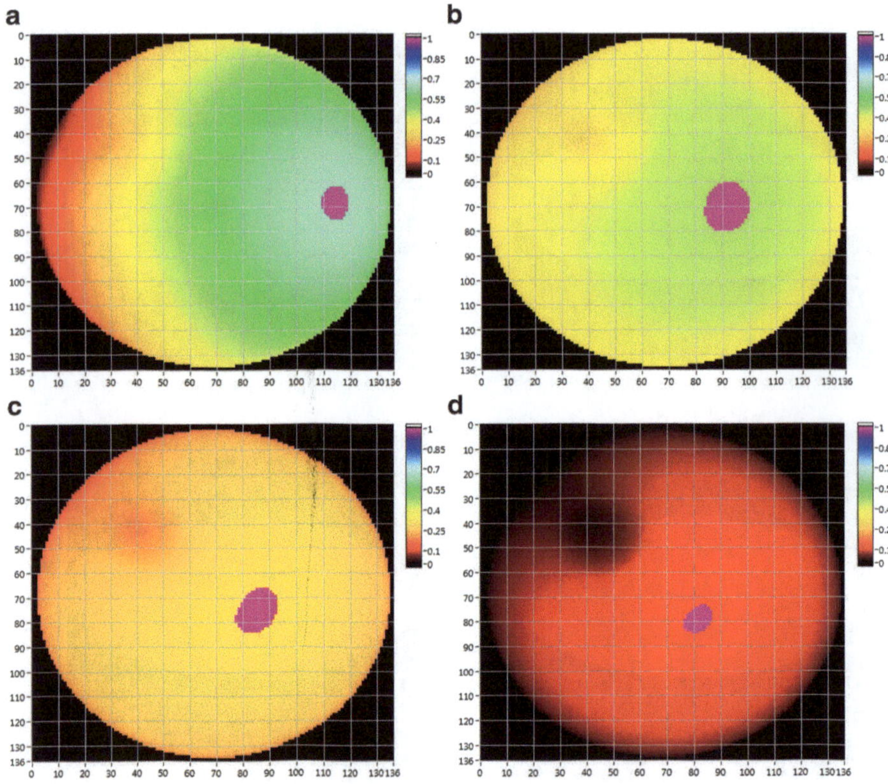

Fig. 5.6 Photon density distributions in the slice of the inhomogeneous conical object at the following times after the injection of the pulse: 0.7 ns (**a**), 2.1 ns (**b**), 3.5 ns (**c**), and 4.9 ns (**d**). Dimensions of the pictures are 102 mm × 102 mm

5.5 Conclusion

The described model and computer simulations were able to identify the following regularities of PDNM movement in a conical object:

1. In all homogeneous cases PDNM moves to the geometric center of the object, regardless of the presence or absence of deformations, and values of the absorption and scattering.
2. In presence of an absorbing inhomogeneity, PDNM moves toward the point that is symmetric to its geometric center, relative to the center of conical object regardless of the presence or absence of deformation.
3. Diffuse transmittance intensity decay in the cases with the undeformed object is slightly higher than in those with deformations. The difference is about 8–10 % for the cases described in here.

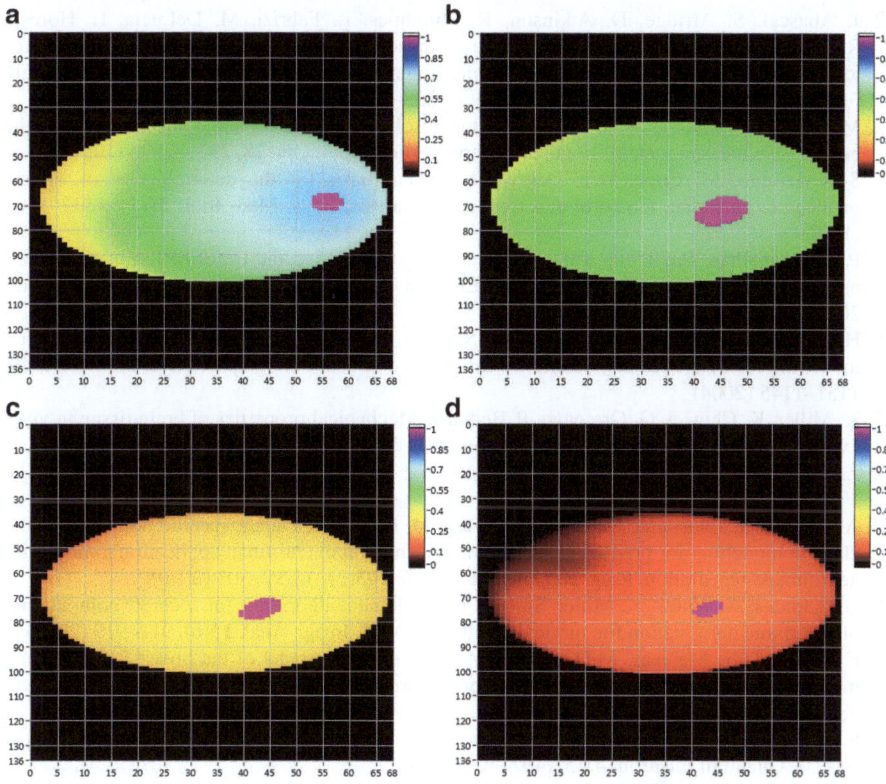

Fig. 5.7 Photon density distributions in the slice of the deformed conical inhomogeneous object at the following times after the injection of the pulse: 0.7 ns (**a**), 2.1 ns (**b**), 3.5 ns (**c**), and 4.9 (**d**). Dimensions of the pictures are 51 mm × 102 mm

The described properties allow assessing influence of deformation along with optical properties of the object on the photon density distribution and attenuation. They will be useful in developing more effective methods for solving DOT inverse problem.

Acknowledgement This work is supported by the State contract in Biomedical diagnostics, grant 12.849.2014/K2014 and the Scholarship of the President of Russian Federation, grant SP-4213.2013.4.

References

1. D.A. Zimnyakov, V.V. Tuchin, Optical tomography of tissues. Quantum Electron. **32**(10), 849–867 (2002)

2. J. Abascal, S. Arridge, D. Atkinson, R. Shindmes, L. Fabrizi, M. DeLucia, L. Horesh, R. Bayford, D. Holder, Neuroimage: use of anisotropic modelling in electrical impedance tomography; description of method and preliminary assessment of utility in imaging brain function in the adult head. Neuroimage **43**(2), 258–268 (2008)
3. M. Patterson, B. Chance, B. Wilson, Time resolved reflectance and transmittance for the noninvasive measurement of tissue optical properties. Appl. Opt. **28**, 2331–2336 (1989)
4. S.R. Arridge, M. Cope, D.T. Delpy, Theoretical basis for the determination of optical pathlengths in tissue: temporal and frequency analysis. Phys. Med. Biol. **37**(7), 1532–1560 (1992)
5. P. Pathmanathan, Predicting Tumour Location by Simulating the Deformation of the Breast using Nonlinear Elasticity and the Finite Element Method. PhD thesis. University of Oxford, 2017 (2006)
6. H. Dehghani, M.M. Doyley, B.W. Pogue, S. Jiang, J. Geng, K.D. Paulsen, Breast deformation modelling for image reconstruction in near infrared optical tomography. Phys. Med. Biol. **49**, 1131–1145 (2004)
7. K. Miller, K. Chinzei, G. Orssengo, P. Bednarz, Mechanical properties of brain tissue in vivo: experiment and computer simulation. J. Biomech. **33**(11), 1369–1376 (2000)
8. Z. Wang, Mechanical and Optical Methods for Breast Cancer Imaging. PhD thesis. The University of Iowa, 2010, p. 184
9. A. Wittek, G. Joldes, M. Couton, S.K. Warfield, K. Miller, Patient-specific non-linear finite element modelling for predicting soft organ deformation in real-time; application to non-rigid neuroimage registration. Prog. Biophys. Mol. Biol. **103**(2–3), 292–303 (2010)
10. G.R. Joldes, A. Wittek, K. Miller, Suite of finite element algorithms for accurate computation of soft tissue deformation for surgical simulation. Med. Image Anal. **13**(6), 912–919 (2009)
11. F.S. Azar, D.N. Metaxas, M.D. Schnall, A finite element model of the breast for predicting mechanical deformations during biopsy procedures, in *Proceedings of the IEEE*, 2000, pp. 38–45
12. S.G. Proskurin, Using late arriving photons for diffuse optical tomography of biological objects. Quantum Electron. **41**(5), 402–406 (2011)
13. S.G. Proskurin, A.Y. Potlov, Early- and late-arriving photons in diffuse optical tomography. Photonics Laser Med. **2**(2), 139–145 (2013)
14. A.Y. Potlov, S.G. Proskurin, S.V. Frolov, Three-dimensional representation of late-arriving photons for detecting inhomogeneities in diffuse optical tomography. Quantum Electron. **44**(2), 174–181 (2014)
15. D. Durian, J.J. Rudnick, Spatially resolved backscattering: implementation of extrapolation boundary condition and exponential source. J. Opt. Soc. Am. A **16**(4), 837–844 (1999)
16. M. Schweiger, S.R. Arridge, Application of temporal filters to time-resolved data in optical tomography. Phys. Med. Biol. **44**, 1699–1717 (1999)
17. H. Dehghani, S. Srinivasan, B. Pogue, A. Gibson, Numerical modelling and image reconstruction in diffuse optical tomography. Phil. Trans. R. Soc. A **367**, 3073–3093 (2009)
18. A.Y. Potlov, S.V. Frolov, S.G. Proskurin, Movement of the photon density normalized maximum in homogeneous and inhomogeneous media with tissue-like optical properties. Laser Phys. **25**, 035601 (2015)
19. S.G. Proskurin, A.Y. Potlov, S.V. Frolov, Specific features of diffuse photon migration in highly scattering media with optical properties of biological tissues. Quantum Electron. **45**(6), 540 (2015)
20. A.B. Konovalov, V.V. Vlasov, A.G. Kalintsev, O.V. Kravtsenyuk, V.V. Lyubimov, Time-domain diffuse optical tomography using analytic statistical characteristics of photon trajectories. Quantum Electron. **36**(11), 1048–1055 (2006)

Chapter 6
Mechanical Models of Endothelial Mechanotransmission Based on a Population of Cells

Yi Chung Lim, Michael T. Cooling, Sue R. McGlashan, and David S. Long

6.1 Introduction

Endothelial cells (ECs) detect and respond to blood flow-induced forces in a process known as mechanotransduction. Dysfunctional mechanotransduction has been implicated as one of the causes of cardiovascular diseases such as atherosclerosis [1]. Hence, studying mechanotransduction processes is motivated by the possibility of improving the diagnosis and prevention of cardiovascular diseases. Davies et al. proposed a decentralized model in which mechanotransduction occurs as the sum of two processes. First, mechanotransmission, whereby forces from blood flow are transmitted internally into the cell via the cytoskeleton. Second transduction, where force-sensitive transducers located throughout the cell are activated, causing a biological response such as release of vasodilators [2]. Hence there is no central transducer responsible for overall mechanotransduction: cell signaling is the aggregate output of all transducers within the cell.

Both mechanotransmission within and mechanical behavior of endothelial cells have been examined using computational modeling [3–6]. The spatial domain of these studies is typically based either on images of a single cell [5] or on idealized geometry [3, 4, 6], where the cell is represented by simple shapes and solids (for instance, the nucleus as a spheroid). These approaches do not consider the effect of cell shape variation on their predicted outputs. As such, it is uncertain whether the findings of cell-specific studies can be applied to the overall cell population.

There is reason to suggest they cannot be: Ferko et al. demonstrated that the spatial distribution of focal adhesions causes heterogenous stress/strain distributions. They also found that stresses concentrated at the interface of the nucleus and cytoplasm [5]. This suggests that varying focal adhesion and nuclei morphology

Y.C. Lim • M.T. Cooling • S.R. McGlashan • D.S. Long (✉)
Auckland Bioengineering Institute, University of Auckland, Auckland, New Zealand
e-mail: d.long@aucklanduni.ac.nz

G. Joldes et al. (eds.), *Computational Biomechanics for Medicine*,
DOI 10.1007/978-3-319-28329-6_6

would have resulted in substantially different stress/strain estimates. Caile et al. found that rounded cells had an elastic response to compression, whereas spread cells of identical elastic moduli exhibited hysteresis [3]. Taken together, this suggests that individual endothelial morphology is an important determinant of cell mechanical behavior. Thus it is important to consider spatial variation in the population of cells when performing single cell mechanical simulations.

In this study we aimed to quantify the morphological variation in an EC population. We also aimed to determine if this variation leads to substantially different estimates of mechanical behavior. The starting point for our methods to quantify spatial variation was based on a range of studies carried out by Murphy et al. [7–9]. They proposed a generative model approach, whereby a population of cells are imaged and used to build up a dataset of spatial descriptors. From these descriptors, new virtual cells can be generated that display characteristic patterns learned from the cell images.

6.2 Materials and Methods

We imaged a population of endothelial cells cultured under identical conditions ($n = 15$). The nucleus, f-actin, and acetylated α-tubulin components of the cytoskeleton were imaged. Shape descriptors were formulated to numerically describe the morphology of the nucleus and cell edge. The statistical variation in each descriptor was analyzed and used to create generative cells: virtual cells with a morphology sampled from the shape descriptor distribution of the entire population of endothelial cells. These generative cells were then used as the spatial domain for our finite-element analysis to generate stress estimates. Physiological levels of flow-induced shear stress formed the boundary conditions of our simulation.

6.2.1 Cell Culture and Imaging

Unless otherwise stated all materials were obtained from Life Technologies (Carlsbad, CA, USA). Human microvascular endothelial cells (HMEC-1s) were kindly provided by Dr. Edwin Ades, Mr. Francisco J. Candal (CDC, Atlanta GA, USA), and Dr. Thomas Lawley (Emory University, Atlanta, GA, USA) [10]. HMEC-1s between passages 5–7 were seeded (1×10^5 cells/ml concentration) onto fibronectin-coated 6-well plates (fibronectin, 20 μg/ml, #33016-015). They were grown to confluence at 37 °C in 5 % CO_2. Cells were maintained daily in MCDB131 (#10372019) media with 10 % L-glutamine (#25030081), 2 % FBS (#10091148), and 1 % penicillin/streptomycin (#15140122).

Immunofluorescent Labeling Due to overlapping wavelengths of absorption and emission only three distinct sub-cellular components could be imaged simultaneously. Thus we stained and imaged the nucleus, acetylated α-tubulin, and f-actin. The protocol is as follows:

Upon confluence cells were fixed with 4 % paraformaldehyde (#158127-100, Sigma-Aldrich, St Louis, MO, USA) for 30 min at 37 °C, then washed with PBS (3 × 5 min, #00-3000). Next, cells were permeabilized in triton X-100 (0.5 %, 5 min, #T9284, Sigma-Aldrich), followed by PBS wash (3 × 5 min). To image the nucleus, cells were stained with Hoechst 33258 (1:1000, #B2883, Sigma-Aldrich) for 5 min at room temperature and washed with PBS (3 × 5 min). To image acetylated α-tubulin, cells were blocked with goat serum (1:20, #G9023 Sigma-Aldrich) for 30 min at room temperature. They were then incubated overnight with 611b (1:500, #T7451 Sigma-Aldrich). This was followed by a 2 h incubation with secondary antibody goat antimouse Alexa Fluor 594 (1:500, #A11005), and a PBS wash (3 × 5 min). To image f-actin, cells were incubated with Alexa Fluor Phalloidin 488 (1:500, #A12379) for 30 min at room temperature, followed by PBS washes (3 × 10 min). Next, coverslips were mounted directly onto six well plates using ProLong Gold (#P36934). The bottom of each well (with coverslip attached) was then removed with a heated scalpel to allow direct imaging.

Microscopy An Olympus FV1000 laser scanning confocal microscope with a 60×/1.35 NA oil immersion lens was used to image the cells. Diode-pumped 405 nm (to image the nuclei), helium neon 543 nm (acetylated α-tubulin), and an argon ion multiline 458 nm (primary cilium and f-actin) lasers were used to sequentially excite samples. Acquired image resolution was 1600 × 1600 pixels, with an XY spatial resolution 0.132 μm/pixel.

6.2.2 Generating Virtual Cell Components

Image processing and image analysis of the nucleus and cell edge were carried out in MATLAB (version R2013b), ImageJ (version 1.48o), and AMIRA (version 5.6). Nuclei morphology was quantified first, as the nucleus is an easily identified feature present in every cell. Thus the nucleus functions as a useful reference point for shape descriptors of the other sub-cellular components.

Quantifying Spatial Variation of Nuclei To quantify nuclei morphology we used a modified version of the method described by Buck et al. [9]. This method involved four spatial descriptors: median axis, nucleus width, nucleus length, and centroid vector (see Fig. 6.1). Both median axis and nuclei curves were normalized by the length of the central axis, and fitted with a 10th order polynomial. Finally in every 2D slice, we defined the centroid vector: the vector between the centroid of the slice and the centroid of the whole nucleus. Hence any nucleus can be described by the $1 \times n$ vector of central axis lengths where n is number of slices, two $11 \times n$

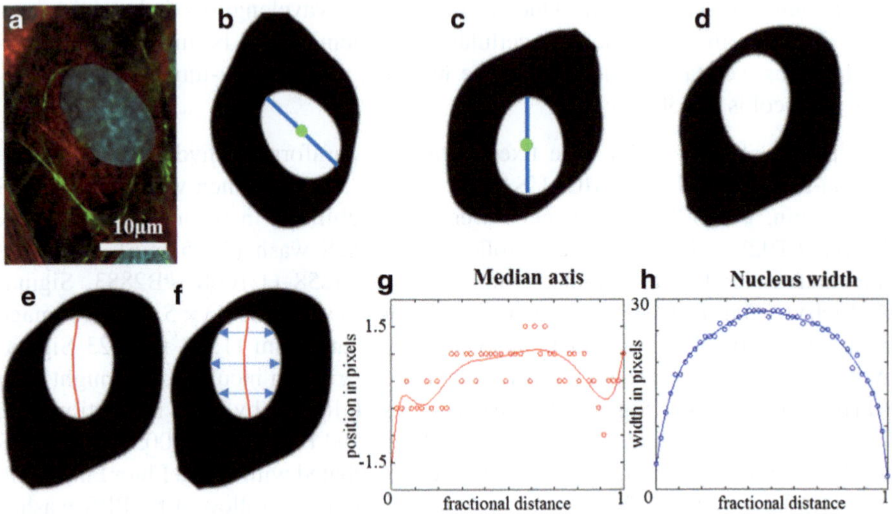

Fig. 6.1 Quantifying nuclei morphology using spatial descriptors. (**a**) Triple-labeled co-image of a human microvascular endothelial cell, with the nucleus in *blue*, f-actin in *green*, and acetylated α-tubulin in *red*. (**b**) Thresholded image of cell shown in (**a**) with the central axis in *blue* and the nucleus centroid in *green*. The central axis length is the nucleus length. (**c**) Cells were rotated so that the central axis runs top-to-bottom. (**d**) Cells were rotated a 180° if needed, to ensure majority of cell area was on the right-hand side of the central axis. All slices in stack were rotated by the same angle as the central slice. (**e**) The median axis was found, as the point along the row (shared *y* coordinate) that is equally distant from either edge of the nucleus. (**f**) This distance is known as the nucleus width. (**g**) Median axis position in pixels, normalized by dividing by nuclear length. (**h**) Nuclear width in pixels normalized by dividing by nuclear length

vectors (fitted coefficients of a 10th order polynomial to describe the median axis and nuclei width), and the $3 \times n$ array of centroid vectors. A distribution for each shape descriptor was gathered from a population of cells ($n = 15$). By sampling from this distribution, it is possible to generate a nuclei representative of the entire population.

Quantifying Spatial Variation of Cell Edge The cell membrane was not explicitly imaged in this study. Instead the edge of the cell was approximated as being one pixel (132 nm) beyond the edge of the f-actin and acetylated α-tubulin features. This was achieved by adding these channels together and Gaussian blurring the resulting image. The shape descriptors of the cell membrane were adapted from Buck et al. [9]. First, the nuclear centroid of the central slice within the cell was defined as the origin of the cell. Every cell image stack was rotated so that the nucleus central axis ran top-to-bottom and majority of cell area was on the right-hand side of the central axis (Fig. 6.1b–d). The centroid of each slice was also determined, and the x and y displacement between the slice centroid and the origin was recorded. In each slice the cell boundary was detected by finding the boundary at 240 equally spaced points, radiating outwards 1.5° apart from the slice centroid (see Fig. 6.4).

Instead of storing these points as x, y, and z coordinates, they were converted to a polar coordinate system, hence only 240 radial lengths were needed (as the angle is known).

Hence the cell membrane spatial description could be stored as a $241 \times n$ cell edge vector, where n is the number of slices, and there are 240 radial lengths, and a single z coordinate that all the points in each slice share. Prior to calculating radial length, each x and y coordinate in the cell edge vector was normalized by the displacement of the slice centroid relative to the origin (Fig. 6.3).

To analyze the variation in these cell edge vectors, we used principle component analysis. Firstly the $241 \times n$ cell edge vectors were converted back into Cartesian coordinates resulting in a vector of $1 \times 720n$, whereby each slice, n, has 240 x, y, and z coordinates. A matrix was formed from the edge vectors of all cells, $15 \times 720n$. Each column was centered by subtracting the mean cell edge vector. PCA was then performed using singular value decomposition algorithm. The PCA method finds shape modes that can be linearly combined to recreate any cell shape from the original data.

Hence cell membrane shapes can be generated by randomly generating shape-mode weightings. The square root of the eigenvalue is the standard deviation of that particular shape mode (corresponding eigenvector) in the population. Hence the weightings can be sampled from a normal distribution with a mean (zero, as data is centered), and standard deviation, to generate "typical" cells, or sampled from a standard deviation above or below the mean to generate "unlikely" cells.

To deconstruct our actual cell shapes into a linear combination of shape modes, we used the Open Genetic Algorithm Toolbox, implemented in MATLAB [11]. The parent solutions of the genetic algorithm were a 1×8 vector of shape-mode weightings. The fitness function of the algorithm was to minimize the root-mean-squared difference between the 240 points in the original cell and the 240 points of the cell created from the genetic algorithm. Rank scaling method and satellite range scheduling selection method were used, with a single crossover point, crossover probability of 90 %, mutation probability of 6–9 %, and an elitism of 10 %. Note: the genetic algorithm parameters reported here were used in this study to analyze all the cells. However, the genetic algorithm converged to a similar solution when top scaling was used and when the elitism and mutation rate was varied.

6.2.3 Finite-Element Model of Solid and Fluid Domains

Computational modeling was carried out in ANSYS (version 16), using geometry files processed in SolidWorks (version 2011).

Fluid Domain We simulated a single cell within a flow chamber. Flow inlet and outlets were 300 μm up- and downstream of the cell. The side and upper walls were 200 μm away from the cell. This geometry approximates the flow chamber we intend to use in future experiments. Pressure boundary conditions of 6 and 0 Pa

Table 6.1 Constitutive properties of computational model

Parameter	Value	Reference
Poisson's ratio, nucleus	0.33	[3, 5]
Poisson's ratio, cytoplasm	0.33	[3, 5]
Young's modulus, nucleus	5100 Pa	[3–6, 17]
Young's modulus, cytoplasm	775 Pa	[3–6, 17]

were applied to the inlet and outlet, respectively, resulting in a pressure drop of 0.009 Pa/μm. This resulted in a maximum fluid velocity of 4.7×10^{-2} ms^{-1}, and a maximum wall shear stress of 2.4 Pa on the cell surface and $\sim$0.8 Pa on the chamber walls. Our boundary conditions were selected to result in a similar wall shear stress as simulated by Ferko et al. (1 Pa wall shear stress on the chamber walls with $\sim$1.5 Pa maximum wall shear stress on the apical cell surface) [5]. The Reynolds number was $\sim$9, indicating laminar flow. The forces acting on the cell surface were exported from the fluid domain into the solid domain (one-way fluid–structure interaction).

Solid Domain The basal surface of the cell was constrained in all directions, simulating cell attachment with the extracellular matrix. We treated both the nucleus and the cytoplasm as compressible isotropic linear elastic materials, as assumed by Ferko et al. [5]. The full list of material properties is shown in Table 6.1.

Meshing and Solution Procedure The fluid domain was meshed with 4-node tetrahedron elements. The solid domain was meshed with 10-node ANSYS solid 187 elements (4 vertices, 6 mid-edge nodes). Both fluid and solid domains were meshed using the patch conforming method, which firstly generates a surface mesh, then uses the Delaunay advancing front approach to mesh the remaining volume. Mesh independence analysis was conducted on both the solid and fluid domain (see Fig. 6.2).

6.3 Results

6.3.1 Shape Variation of Endothelial Cell Components

Nucleus The median axis and nucleus width are shown in Fig. 6.3. The mean centroid vector was 0 ± 1.5 pixels in both x and y, indicating that each slice of the nucleus had a centroid directly above the middle slice centroid. The average nuclear length was 144 ± 20.4 pixels. By sampling from these distributions it was possible to generate a typical nucleus as shown in Sect. 6.3.2.

Cell Edge Each cell in the population was deconstructed into a linear combination of shape modes. Of the 15 cells, 5 had non-typical morphology. They had a first shape modes weighting greater than one standard deviation away from the average (the first shape mode explains 40 % of the shape variation, Fig. 6.4).

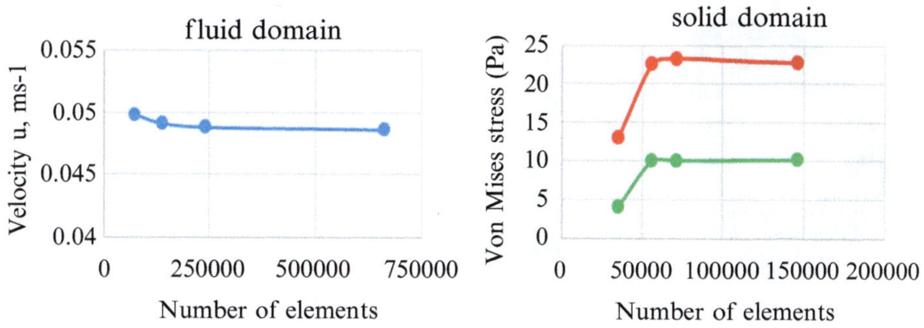

Fig. 6.2 Mesh independence analysis of the fluid (*left*) and solid domain (*right*). The fluid domain solution stabilized at 1.4×10^5 number of elements. The plotted point is the velocity in the middle of the channel 100 μm downstream of the cell. Because the system Reynolds number is ~9, we can approximate the system using a numerical solution to laminar flow in a rectangular duct, as described by Spiga and Morino [16]. We estimate a maximum velocity of 0.50 ms^{-1} which is in close agreement with our simulated value of 0.49 ms^{-1}. To analyze the solid domain we monitored the Von Mises stress at two points: a point on the apical surface of the nucleus near the maximum stress concentration (*red*) and a point on the basal surface of the nucleus (*green*). The solid domain solution stabilized at 5.7×10^4 number of elements with an element sizing of 2.5 μm. This sizing was used in both the typical and atypical cell models

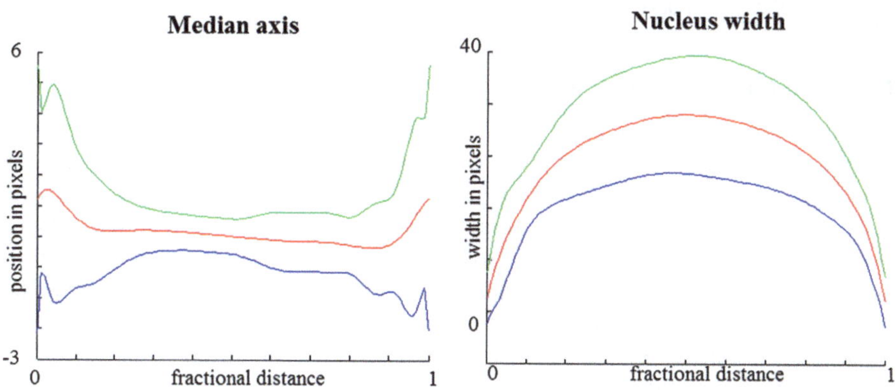

Fig. 6.3 Variation in nuclei shape descriptors. The average median axis and nucleus width are shown in *red*. The *green* and *blue lines* represent an envelope of two standard deviations above and below the average, hence nearly all nuclei fall within the envelope. The maximum average width of the nucleus is at the center, and is 40 pixels × 0.132 μm/pixel resolution = 5.28 μm (on each side of the median axis)

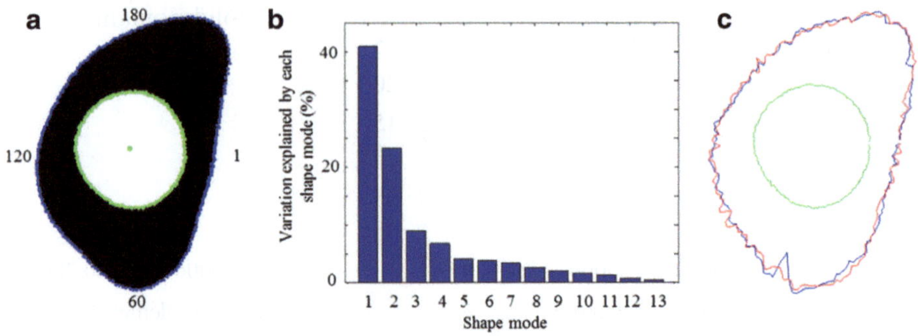

Fig. 6.4 Spatial descriptors of the cell edge. (**a**) Illustration of polar coordinate description of the cell outline. There are 240 data points shown on the outline in *blue*, numbers refer to the point number. Hence each slice can be represented by a 1 × 241 vector. (**b**) Percentage of shape variation explained by each of the shape modes found using PCA. To analyze our data set we considered the first eight modes which accounted for >95 % of the total variation. (**c**) Goodness of fit of the genetic algorithm (*blue*) versus the actual cell outline (*red*). The actual nucleus is shown in *green*. To generate the *blue* cell outline the weightings found using the genetic algorithm were multiplied with the shape modes and added to the average cell outline. PCA analysis and genetic algorithm fitting have been extended to three dimensions, but for illustrative purposes are shown here in two dimensions

6.3.2 Computational Model Estimates of Stress Depend on Cell Morphology

6.4 Discussion and Conclusions

Overall we found littlevariation in nuclear shape in the population: both the median axis and median width curves are symmetric, and the nuclear centroid vector is zero, suggesting the nucleus has three perpendicular planes of symmetry. Thus, overall size is the main component of morphological variation in the nucleus of static endothelial cells. The size variation of the nucleus is relatively small compared to size variation in the overall cell. Taken together, this suggests that computational mechanical estimates of the isolated nucleus can be generalized as there is little morphological variation.

We found that there was significant morphological variation in overall cell shape. If one of the five non-typical morphology cells had been selected for a cell-specific analysis, the findings could not be generalized.

Our estimated stress findings were in agreement with that of Ferko et al. (<60 Pa) [5]. Even with a simple finite-element analysis (linear, elastic, homogenous, and compressible) we found substantial differences in the stress distribution between the typical and non-typical morphology cell models (Fig. 6.5). The trend in model development is towards increasing sophistication and the inclusion of more discrete

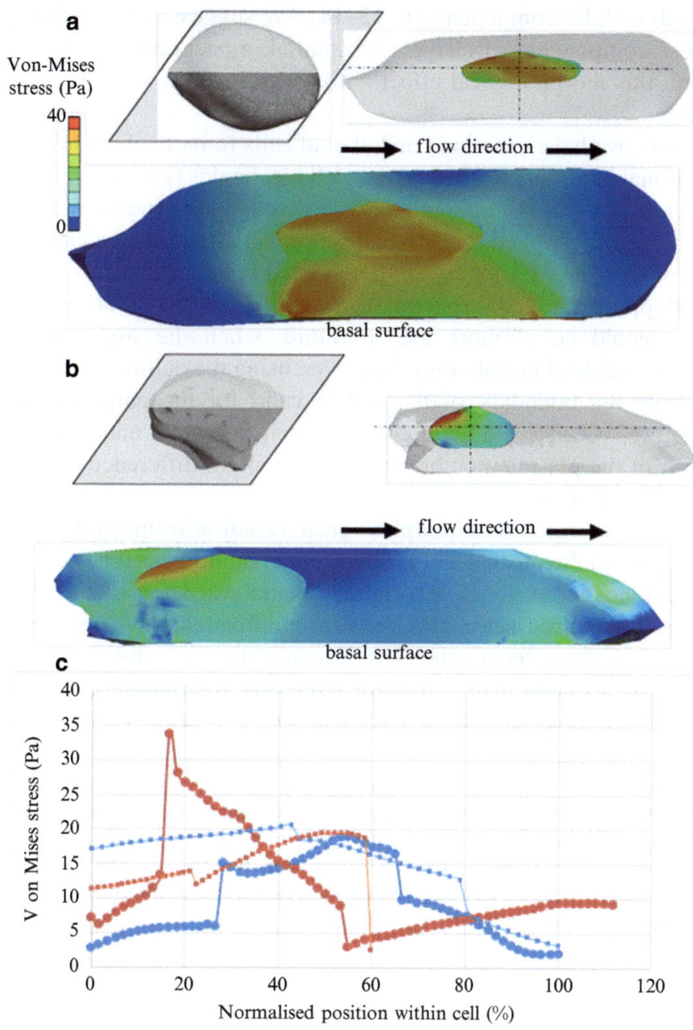

Fig. 6.5 Comparison of Von Mises stress estimates in typical versus non-typical morphology. (**a**) Synthetic "typical" cell with the first eight shape modes within one standard deviation of the average. Cross-sectional plane is indicated, as is nuclei position within the cell. Stresses are evenly distributed within the nucleus, and concentrate on the upstream side below the nuclei. (**b**) Synthetic "non-typical" cell with the first four shape modes between one and two standard deviations away from the average. Stress is concentrated on the upstream side of the nucleus, and above the nucleus (on upstream side). We no longer observe stress concentration below the nucleus. To isolate the effect of different cell edge morphology and nuclei position within the cell, the same nucleus was used in both models. Identical computational parameters were used for both models with the exception of the spatial domain (the same minimum element lengths, simulated flow, etc.). (**c**) Von Mises stress along the dotted lines shown in (**a, b**). Typical cell is in *blue*, atypical in *red*; *thick lines* with round data points correspond to the path parallel to flow direction. *Thin lines with square data* points correspond to basal–apical axis path. Each path has been normalized to the typical cell's length (upstream to downstream, basal surface to apical). Nuclei boundaries cause sudden jumps in the Von Mises stress profile. Stresses are symmetrical upstream and downstream of the nuclei in the typical cell, but are highly asymmetrical in the atypical cell

additional sub-cellular components [6, 12–14]. We suggest that morphological variation of these components will have a substantial impact on mechanical estimates. Furthermore, this morphological effect is likely to increase with increasing model sophistication.

In this study, we have examined endothelial cells in particular. However, the use of computational modeling to characterize cell mechanics is common in a number of other cell types [12–14]. The population-based shape description methods detailed here could be readily applied to these cell types, in particular, to adherent cells.

The cell membrane was not explicitly modeled in our analysis: given limitations of the overlapping antibody spectra, it was decided that imaging cytoskeletal components would be of more use in future, when the study is extended to incorporate cytoskeletal morphology. However, using the outline of the cytoskeleton to approximate the boundary of the cell is valid for the purposes of computational modeling, because the actin cortex is rich in f-actin and is located within 128 ± 28 nm of the plasma membrane [15]. The slight difference is accounted for by our Gaussian blurring.

We have demonstrated how morphological variation in the cell membrane has significant effect on the mechanical estimates of endothelial cell behavior. In future, we aim to extend our study to incorporate spatial variation of the cytoskeleton: including alpha-tubulin, f-actin, and intermediate filaments. We also aim to extend our study to incorporate focal adhesions. Because they have the function to adhere the cell to the extracellular matrix, the size and spatial distribution of focal adhesions directly affect computational estimates of endothelial mechanics [5].

Acknowledgments Yi Chung Lim is supported by a University of Auckland Doctoral Scholarship. This work was supported by a Faculty Research Development Fund grant (3702516, D.S.L.). We thank Ms. Hilary Holloway and Ms. Jacqui Ross from the Biomedical Imaging Research Unit for assistance in microscope training and image acquisition. Finally, we thank Dr. Edwin Ades and Mr. Francisco J. Candal of CDC and Dr. Thomas Lawley of Emory University for developing the HMEC-1 line and providing it to us (NCEZID-R147589-00).

References

1. P.F. Davies, C. Shi, N. DePaola, B.P. Helmke, D.C. Polacek, Hemodynamics and the focal origin of atherosclerosis. Ann. N. Y. Acad. Sci. **947**(1), 7–17 (2001)
2. P.F. Davies, Hemodynamic shear stress and the endothelium in cardiovascular pathophysiology. Nat. Clin. Pract. Cardiovasc. Med. **6**(1), 16–26 (2009)
3. N. Caille, O. Thoumine, Y. Tardy, J. Meister, Contribution of the nucleus to the mechanical properties of endothelial cells. J. Biomech. **35**(2), 177–187 (2002)
4. R.P. Jean, C.S. Chen, A.A. Spector, Finite-element analysis of the adhesion-cytoskeleton-nucleus mechanotransduction pathway during endothelial cell rounding: axisymmetric model. J. Biomech. Eng. **127**(4), 594–600 (2005)
5. M. Ferko, A. Bhatnagar, M. Garcia, P. Butler, Finite-element stress analysis of a multicomponent model of sheared and focally-adhered endothelial cells. Ann. Biomed. Eng. **35**(2), 208–223 (2007)

6. M. Dabagh, P. Jalali, P.J. Butler, J.M. Tarbell, Shear-induced force transmission in a multicomponent, multicell model of the endothelium. J. R. Soc. Interface **11**(98), 20140431 (2014)
7. T. Zhao, R.F. Murphy, Automated learning of generative models for subcellular location: building blocks for systems biology. Cytometry A **71**(12), 978–990 (2007)
8. T. Peng, R.F. Murphy, Image-derived, three-dimensional generative models of cellular organization. Cytometry A **79**(5), 383–391 (2011)
9. T.E. Buck, J. Li, G.K. Rohde, R.F. Murphy, Toward the virtual cell: automated approaches to building models of subcellular organization "learned" from microscopy images. Bioessays **34**(9), 791–799 (2012)
10. E.W. Ades, F.J. Candal, R.A. Swerlick, V.G. George, S. Summers, D.C. Bosse, T.J. Lawley, HMEC-1: establishment of an immortalized human microvascular endothelial cell line. J. Invest. Dermatol. **99**(6), 683–690 (1992)
11. A.L. Freitas, Open genetic algorithm toolbox. creative commons attribution non-commercial license V2.0 (2012), http://sourceforge.net/projects/gatoolbox/
12. N. Slomka, A. Gefen, Confocal microscopy-based three-dimensional cell-specific modeling for large deformation analyses in cellular mechanics. J. Biomech. **43**(9), 1806–1816 (2010)
13. S. Barreto, C.H. Clausen, C.M. Perrault, D.A. Fletcher, D. Lacroix, A multi-structural single cell model of force-induced interactions of cytoskeletal components. Biomaterials **34**(26), 6119–6126 (2013)
14. H. Khayyeri, S. Barreto, D. Lacroix, Primary cilia mechanics affects cell mechanosensation: a computational study. J. Theor. Biol. **379**, 38–46 (2015)
15. A. Clark, K. Dierkes, E. Paluch, Monitoring actin cortex thickness in live cells. Biophys. J. **105**(3), 570–580 (2013)
16. M. Spiga, G.L. Morino, A symmetric solution for velocity profile in laminar flow through rectangular ducts. Int. Commun. Heat Mass Transfer **21**(4), 469–475 (1994)
17. A.B. Mathur, G.A. Truskey, W. Monty Reichert, Atomic force and total internal reflection fluorescence microscopy for the study of force transmission in endothelial cells. Biophys. J. **78**(4), 1725–1735 (2000)

[illegible]

Chapter 7
Investigation of Modelling Parameters for Finite Element Analysis of MR Elastography

Lyam Hollis, Lauren Thomas-Seale, Noel Conlisk, Neil Roberts, Pankaj Pankaj, and Peter R. Hoskins

7.1 Introduction

Changes in the material properties of tissue are common in many disease states. Indeed in certain conditions such as breast cancer manual palpation to detect a change in material properties has long been used as a method of diagnosis [1]. This method however is only appropriate for superficial abnormalities, dependent on the skill of the individual clinician [2] and is not quantitative. As such the potential benefits of an imaging technique that could quantify material properties are great.

Magnetic resonance elastography (MRE) is an MRI based technique that aims to achieve this [3]. A motion-encoding gradient is utilized to characterize the displacements of shear waves induced into in vivo tissue using an external harmonically oscillating source. Filters are typically applied to remove noise and an inversion algorithm is then used to calculate a map of the material properties from these displacements [4].

Testing and development of MRE post-processing software requires knowledge of the material properties of the test object such that the resulting MRE measurements can be validated. To this end there are a number of commercial phantoms available with quoted stiffness values [5]. Such phantoms are however limited since they do not allow the user variations in geometry or material properties. To overcome this several studies have described creation of in-house phantoms using

L. Hollis (✉) • L. Thomas-Seale • N. Conlisk • P.R. Hoskins
Centre for Cardiovascular Sciences, University of Edinburgh, Edinburgh, UK
e-mail: S1160274@sms.ed.ac.uk

N. Roberts
Clinical Research Imaging Centre, University of Edinburgh, Edinburgh, UK

P. Pankaj
School of Engineering, University of Edinburgh, Edinburgh, UK

G. Joldes et al. (eds.), *Computational Biomechanics for Medicine*,
DOI 10.1007/978-3-319-28329-6_7

agar gel [6, 7]. However this process is also time consuming and generally requires mechanical testing of the material to attain its true properties, a process requiring its destruction.

Use of synthetic in-silico phantoms is a potential solution to these problems. Material properties are defined within the pre-processing stage of finite element analysis (FEA). This enables a direct comparison between the prescribed material property values and the values attained through analysis of the data using the MRE post-processing software. Using FEA in this manner offers the potential to iterate over a large range of geometries and material properties allowing optimization of the MRE methodology and the ability to obtain quantitative data on issues of clinical interest such as minimum lesion size which may be observed on the elastogram.

To date a number of MRE simulation methodologies have been described in detail. Chen et al. [8] used a two-dimensional model to show increased accuracy in simulated shear wavelengths at higher densities and lower shear moduli. Leclerc et al. [9] iteratively altered 3-dimensional FEA model parameters to match with wave propagation in a phantom. Meanwhile Kolipaka et al. [10] compared uniform beam, plate and shell phantoms with 3-d FEA models of the same structures showing good agreement between MRE scans and FEA datasets. The purpose of FEA development here was to create a technique that could validate inversion algorithms. A more clinically driven study by Thomas-Seale et al. [11] investigated idealized atherosclerotic plaques to show frequency dependent wave disruptions in such geometries.

Despite its use in MRE development there has been little research into the influence of modelling parameters on inversion of FEA datasets. This paper aims to investigate the effects of varying boundary conditions, element type and constraints in modelling of MRE.

7.2 Methods

All simulations have been performed using *Abaqus Version 6.10-1 (Dassault Systèmes Simulia Corp., Providence, Rhode Island, USA)*. An explicit method was used with fixed time increments. The size of these increments was derived using the element-by-element stable increment estimator inbuilt to *Abaqus*. Whilst the increment was dependent upon the type of element used and the size of the mesh it was of order 10^{-5}–10^{-6} s for all simulations.

7.2.1 Geometry

The geometry was created using the inbuilt computer aided design (CAD) package in *Abaqus/CAE* (Fig. 7.1). A cylindrical insert with a radius of 10 mm was placed in a cuboid of dimensions $80 \times 80 \times 50$ mm^3 (Fig. 7.2). Harmonic loads of 100 Hz were applied in the z-direction over a nodal region on the upper surface of the model.

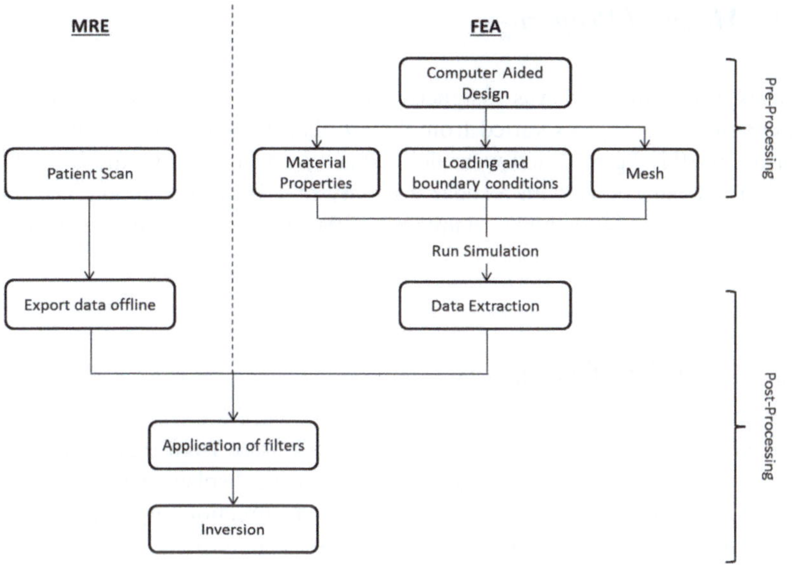

Fig. 7.1 Modelling workflow showing integration of FEA data with MRE post-processing software

Fig. 7.2 The model geometry (**a**), and masks applied to the (**b**) background and (**c**) insert region for calculation of the respective shear modulus values

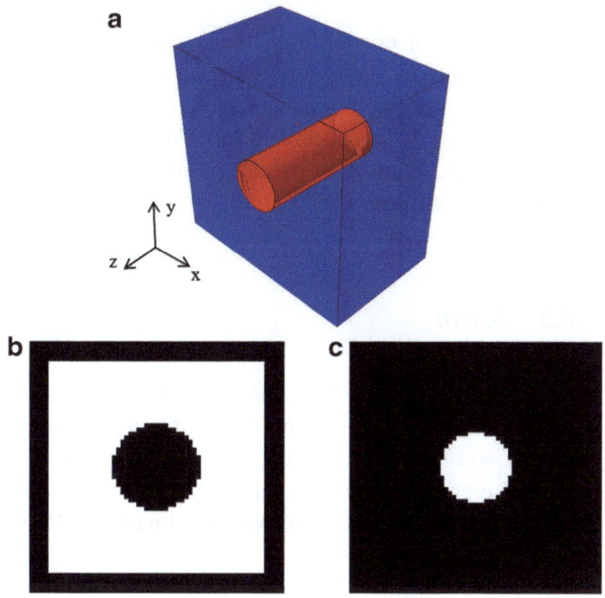

7.2.2 *Material Properties*

The background material was assigned a shear modulus of 3 kPa whilst the shear modulus of the insert was varied from 4 to 9 kPa. Both materials were prescribed a density of 1047 kg m^{-3} and were defined as viscoelastic using the Kelvin–Voigt model of viscoelasticity with a shear viscosity of 1 Pa s throughout. The purpose of this viscosity was to induce damping in the model and reduce the effects of reflections.

7.2.3 *Boundary Conditions*

Boundary conditions were applied to all surfaces aside from that over which the load was applied. Simulations were performed with displacements fixed in each direction independently whilst the remaining two directions were unconstrained, and with displacements fixed in all directions simultaneously.

7.2.4 *Constraints*

Three methods of constraining the two regions of the model to each other were investigated. Firstly the two regions were merged together in the assembly module with the intersecting boundaries retained. Secondly the two regions were tied together using the constraints tool. Finally a frictional interaction was defined between the different regions of the model. Values of 0.5, 0.75, 1 and 1.25 were tested for the coefficient of friction.

7.2.5 *Mesh*

Eight-noded hexahedral (C3D8R), 4-noded linear tetrahedral (C3D4) and 10-noded quadratic tetrahedral elements (C3D10M) were compared. The C3D8R and C3D4 elements were meshed defining a 1 mm element length on all edges. Since the C3D10M elements are of higher order in comparison to the C3D8R and the C3D4 elements a lower mesh density was used with a 1.25 mm element length defined.

7.2.6 *Post-Processing*

Displacements in the *z*-direction were extracted at eight time points 1.25×10^{-3} s apart starting at 0.08 s from the central *xy*-plane with pixel sizes of 1 mm^2. This data was then exported to *Matlab R2013a (MathWorks, Natick, Massachusetts, USA)* where it was reformatted to represent a set of displacement images at each time point. A 1-D Fourier transform was then applied to the data in the temporal direction creating a set of frequency dependent complex wave images. The second of these images corresponds to the frequency of the induced waves and was therefore selected for further analysis. The 2-D direct inversion algorithm was then applied to this image to create a map of the shear modulus, more commonly referred to as an elastogram [12].

Regions of interest were then identified and masks created to isolate these. The areas of these masks were selected avoiding pixels that were within 2 mm of boundaries between the different regions within the model and pixels that were within 5 mm of the edge of the region over which the inversion was performed. The mean value within each region of interest was calculated.

7.2.7 *Convergence Studies*

Convergence studies for each different element type used and convergence assumed to have been achieved when the change in mean value from one increment in mesh density to the next was below 2 % (Fig. 7.3).

7.3 Results

7.3.1 *Boundary Conditions*

Measurements in both the background and the insert were overestimated for all prescribed shear moduli with all the boundary conditions (Fig. 7.4). In the inserts the greatest errors were obtained when displacements were fixed in the *x*-direction, with errors larger than 20 % for all prescribed insert values. Similar overestimations were also shown when displacements were fixed in all directions. The size of the overestimations was greatly reduced when fixed in the *y*- and *z*-directions with errors in the range of 7–13 %.

The size of the overestimations was reduced in the background. The largest overestimation for the majority of the prescribed background values was obtained when displacements were fixed in all directions (7–13 %). The range of error values was similar for all three directions fixed independently (6–9 %).

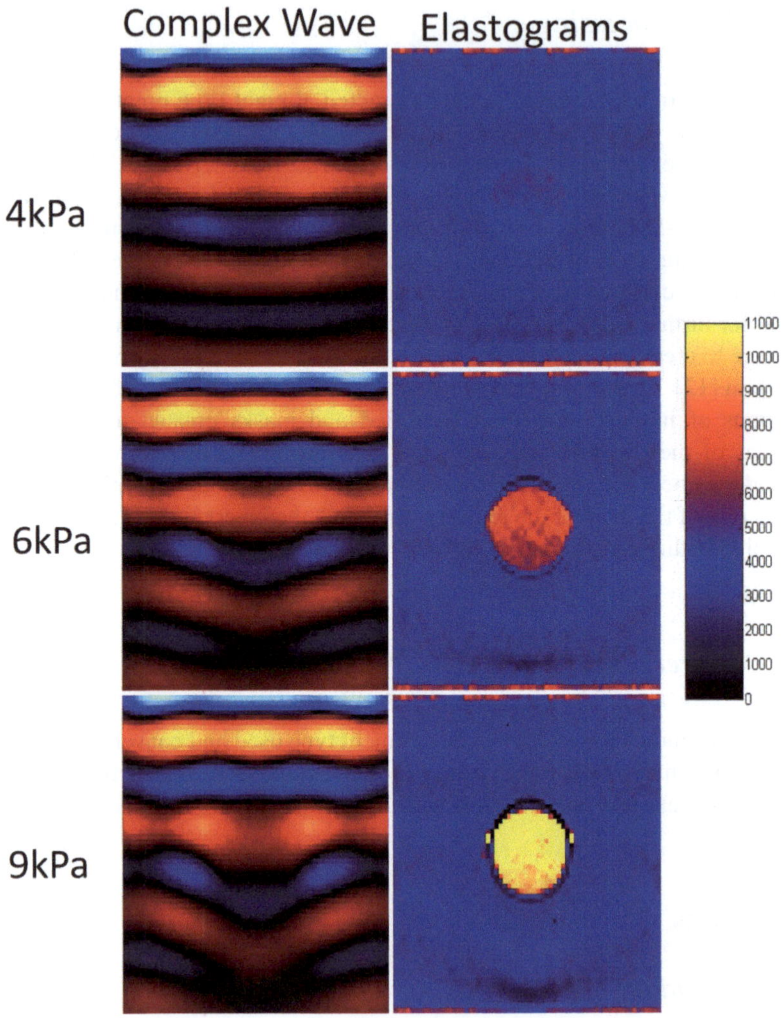

Fig. 7.3 Examples of the complex wave images and respective elastograms from the models prescribed with shear modulus values of 4, 6 and 9 kPa for the insert

7.3.2 Constraints

The background and insert measurements from the tied constraints and the merged parts simulations matched each other almost exactly. Visual inspection of complex wave images showed propagation of the wave in both sections of the model (Fig. 7.5). In all of the frictional interaction simulations the measurements in the background were slightunderestimations of around 2.5 %. Large errors were obtained in the inserts, however, with overestimations typically in excess of 300 %.

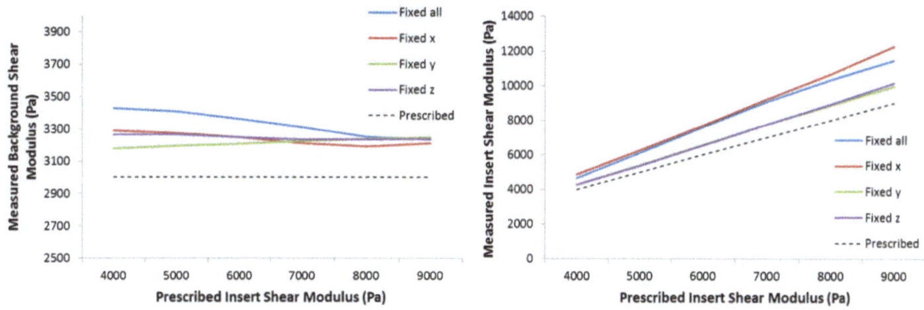

Fig. 7.4 Graphs depicting the shear modulus measurements with the different boundary conditions investigated in this study. The encastred boundary condition leads to large overestimations in both the background and insert regions of the model

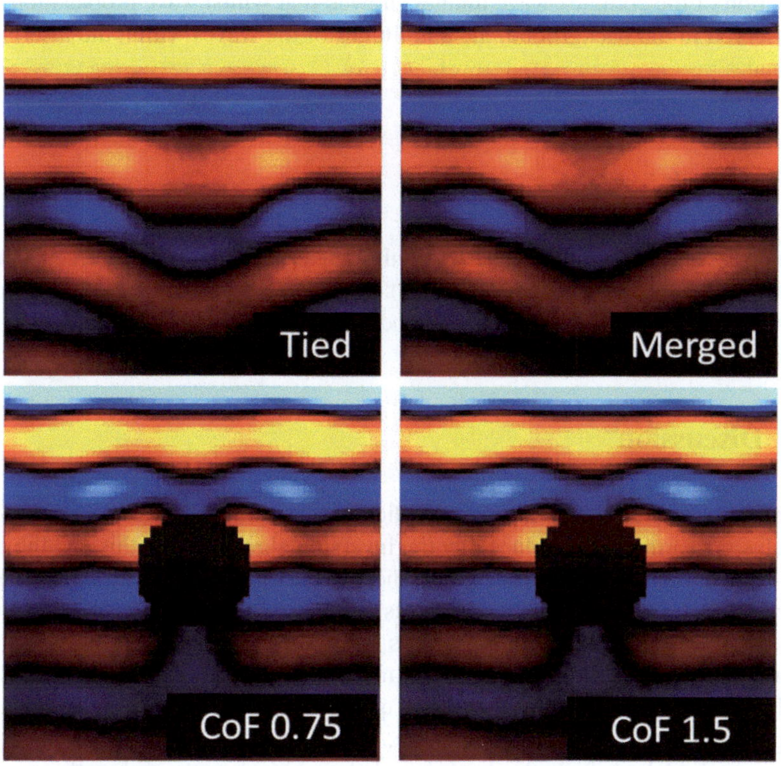

Fig. 7.5 Complex wave images from the different constraint conditions. Whilst there is clear wave propagation when the parts are tied and merged together, wave propagation in the insert is greatly reduced when the constraint is defined using a fractional coefficient

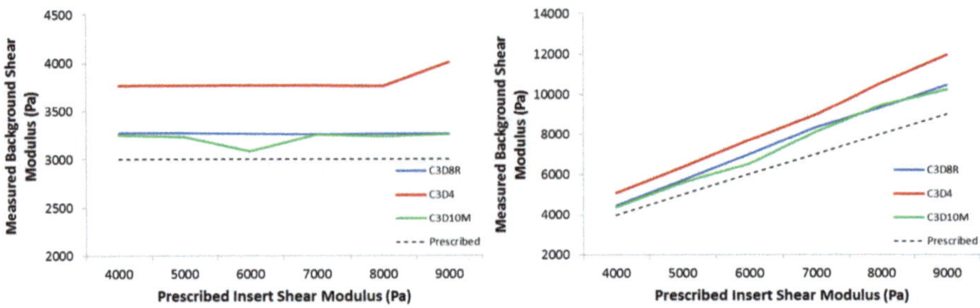

Fig. 7.6 Measured values of the shear modulus for the background and insert regions of the model when meshed with different elements. Whilst the C3D8R and C3D10M elements produce similar measurements, the C3D4 elements result in large overestimations in both regions of the model

Inspection of the complex wave images showed that whilst wave propagation could be observed in the background, the amplitudes of the waves within the inserts were minimal.

7.3.3 Mesh

There was good consistency between the shear modulus measurements in the C3D8R and C3D10M elements (Fig. 7.6). There were large overestimations for the C3D4 elements in both regions of the model (25–30 %).

7.4 Discussion

Modelling parameters can have a significant effect on the accuracy of the shear modulus measurements produced by the inversion algorithm. To this end the use of C3D4 elements dramatically increased the size of overestimations. It is likely that this effect was due to tetrahedral elements being structurally stiffer than hexahedral elements and has important implications when meshing more complex structures that are incompatible with a hexahedral mesh [13, 14]. It was shown here that the problem can be resolved by using the quadratic tetrahedral C3D10M elements, though this increases computational time.

The method by which the boundary between two parts within the model is defined also has an impact in this respect. Whilst the tied condition and merging of parts are unlikely to truly represent the conditions in the human anatomy where sliding of organs with respect to the surrounding tissue is common [15], a frictional interaction prevents transfer of the wave from one region to another. In this paper the resulting lack of wave propagation in the insert leads to large overestimations of the shear modulus.

The choice of boundary condition in MRE modelling is important and has a clear impact upon results. Fixing displacements in all directions has often been used in FEA simulations of MRE [9–11], though in reality the boundaries in MRE will not be totally fixed: waves will pass through a patient into the scan table and with respect to phantoms, the waves will also be transmitted into the phantoms case. Applying a boundary condition where motion is fixed in just one direction and uninterrupted in the other two is also not representative of reality, though also commonly used [8, 16]. Unfortunately the range of boundary conditions offered by *Abaqus*, and indeed most other FEA software packages, is fairly limited typically allowing displacements to be either fixed or unconstrained in each direction. More realistic boundary conditions would take into account the interaction between the model and the surrounding world, though further work is required to determine the nature of such an interaction. The boundary conditions investigated in this study produced a wide range of results with fixing in all directions seeming to result in large overestimations of the stiffness in both the background and insert of the model. To this end fixing in the y-direction alone typically produced the most accurate results in both regions. This was perhaps because motion of the wave in the shear planes was undisturbed when the boundary condition was applied in this direction.

Whilst boundary conditions present a clear problem in the definition of the model and attempting to represent realistic conditions, they also present a problem in the MRE inversion. Firstly the existence of a boundary invalidates the assumption of an infinite homogenous medium upon which the algorithm is derived [17]. Additionally waves reflected off the boundary interfere with those induced and result in interference patterns appearing in the wave image. Whilst reflections and scattering are prevalent in the human body, they are typically more noticeable in phantoms and FEA simulations, where the external boundaries of the model and the boundaries between regions within the model are clearly defined [18, 19]. It is therefore likely that in order to accurately model these regions more complex boundary conditions and constraints than those currently used are required.

7.5 Conclusions

This paper shows the importance of modelling parameters in FEA of MRE. In particular the selection of boundary condition has a significant impact on simulations with large variations between the commonly used conditions used throughout the literature.

References

1. L. Mahoney, A. Crisma, Efficiency of palpation in clinical detection of breast cancer. Can. Med. Assoc. J. **127**(8), 729–30 (1982)
2. M. Tanter, M. Fink, B. Robert, R. Sinkus, B. Larrat, A new rheological model based on fractional derivatives for biological tissues, in *2006 IEEE Ultrasonics Symposium* (IEEE, 2006), pp. 1033–1036

3. R. Muthupillai, D.J. Lomas, P.J. Rossman, J.F. Greenleaf, A. Manduca, R.L. Ehman, Magnetic resonance elastography by direct visualization of propagating acoustic strain waves. Science **269**(5232), 1854–1857 (1995) (New York, NY)
4. A. Manduca, R. Muthupillai, Local wavelength estimation for magnetic resonance elastography, in *International Conference on Image Processing*, vol 3 (1996), pp. 527–530
5. G. Nir, R.S. Sahebjavaher, R. Sinkus, S.E. Salcudean, A framework for optimization-based design of motion encoding in magnetic resonance elastography. Magn. Reson. Med. **00**, n/a–n/a (2014)
6. J.B. Weaver, E.E.W. Van Houten, M.I. Miga, F.E. Kennedy, K.D. Paulsen, Magnetic resonance elastography using 3D gradient echo measurements of steady-state motion. Med. Phys. **28**(8), 1620 (2001)
7. Y.K. Mariappan, K.J. Glaser, R.D. Hubmayr, A. Manduca, R.L. Ehman, K.P. McGee, MR elastography of human lung parenchyma: technical development, theoretical modeling and in vivo validation. J. Magn. Reson. Imaging **33**(6), 1351–61 (2011)
8. Q. Chen, S.I. Ringleb, A. Manduca, R.L. Ehman, K.-N. An, A finite element model for analyzing shear wave propagation observed in magnetic resonance elastography. J. Biomech. **38**(11), 2198–203 (2005)
9. G.E. Leclerc, F. Charleux, M.C. Ho Ba Tho, S.F. Bensamoun, Identification process based on shear wave propagation within a phantom using finite element modelling and magnetic resonance elastography. Comput. Methods Biomech. Biomed. Eng. 37–41 (2013)
10. A. Kolipaka, K.P. McGee, A. Manduca, A.J. Romano, K.J. Glaser, P.A. Araoz, et al., Magnetic resonance elastography: inversions in bounded media. Magn. Reson. Med. **62**(6), 1533–1542 (2009)
11. L. Thomas-Seale, D. Klatt, P. Pankaj, N. Roberts, I. Sack, P. Hoskins, A simulation of the magnetic resonance elastography steady state wave response through idealised atherosclerotic plaques. IAENG Int. J. Comput. Sci. **38**, 394–400 (2011)
12. A. Manduca, T.E. Oliphant, M.A. Dresner, J.L. Mahowald, S.A. Kruse, E. Amromin, et al., Magnetic resonance elastography: non-invasive mapping of tissue elasticity. Med. Image Anal. **5**(4), 237–254 (2001)
13. E. Wang, T. Nelson, R. Rauch, Back to elements-tetrahedra vs. hexahedra, in *Proceedings of the 2004 International ANSYS Conference* (2004)
14. X. Bourdin, X. Trosseille, Comparison of tetrahedral and hexahedral meshes for organ finite element modeling: an application to kidney impact, in *20th Century Enhanced Safety of Vehicles Conference: Innovations for Safety: Applications and Challenges* (Lyon, France, 2007)
15. A. Al-Mayah, J. Moseley, M. Velec, K. Brock, Deformable modeling of human liver with contact surface, in *Science and Technology for Humanity (TIC-STH), 2009 IEEE Toronto International Conference* (2009), pp. 137–140
16. Q. Chen, S. Bensamoun, J.R. Basford, J.M. Thompson, K.-N. An, Identification and quantification of myofascial taut bands with magnetic resonance elastography. Arch. Phys. Med. Rehabil. **88**(12), 1658–61 (2007)
17. S. Papazoglou, U. Hamhaber, J. Braun, I. Sack, Algebraic Helmholtz inversion in planar magnetic resonance elastography. Phys. Med. Biol. **53**(12), 3147–58 (2008)
18. S. Papazoglou, U. Hamhaber, J. Braun, I. Sack, Horizontal shear wave scattering from a nonwelded interface observed by magnetic resonance elastography. Phys. Med. Biol. **52**(3), 675–84 (2007)
19. B.N. Li, C.K. Chui, S.H. Ong, T. Numano, T. Washio, K. Homma, et al., Modeling shear modulus distribution in magnetic resonance elastography with piecewise constant level sets. Magn. Reson. Imaging **30**(3), 390–401 (2012)

Chapter 8
Fuzzy Tissue Classification for Non-Linear Patient-Specific Biomechanical Models for Whole-Body Image Registration

Mao Li, Adam Wittek, Grand R. Joldes, and Karol Miller

8.1 Introduction

Radiographic image registration is important for disease diagnosis, treatment assessment and surgery planning [1–3]. In recent years, numerous image processing-based registration methods have been successfully developed [3–5]. However, many of those methods were proven to be effective for selected body segment, such as the brain, the breast, lungs and prostate [2, 6, 7]. Problems involving large differences between the source and target images (i.e. whole-body CT images) still remain a challenge. Therefore, biomechanical modelling, in which predicting the deformation of organs/tissues is treated as a computational problem of solid mechanics, has been recommended by many researchers [8–10].

The finite element (FE) method has been historically used to predict deformations of body organs/tissues [11–13]. However, creating a patient-specific finite element model is a time-consuming process which involves image segmentation, mesh generation and material property assignment [14, 15]. Despite substantial research effort, automatic medical image segmentation remains an unsolved problem [16]; in particular, when anatomical features depicted in the images are affected by disease/pathology and boundaries between different tissues and organs are difficult to distinguish (which is the case for whole-body CT images and abdominal

M. Li • A. Wittek (✉) • G.R. Joldes
Intelligent Systems for Medicine Laboratory, The University of Western Australia, Perth, Australia
e-mail: adam.wittek@uwa.edu.au

K. Miller
Intelligent Systems for Medicine Laboratory, The University of Western Australia, Perth, Australia

Institute of Mechanics and Advanced Materials, Cardiff School of Engineering, Cardiff University, Cardiff, Wales, UK

G. Joldes et al. (eds.), *Computational Biomechanics for Medicine*,
DOI 10.1007/978-3-319-28329-6_8

organs). Determining patient-specific constitutive properties of the human body tissues is another challenge; despite recent progress in magnetic resonance (MR) and ultrasound elastography [17], reliable non-invasive method to determine such properties in-vivo has not been created yet [18].

In our previous studies, the Fuzzy C-Means (FCM) cluster algorithm was proposed to assign material properties at integration points of the computational grid directly from medical images, without time-consuming image segmentation [19]. As a further development, we applied the FCM algorithm to create a patient-specific whole-body model for computing deformations of body organs/tissues for whole-body CT image registration [20, 21]. The FCM determines the cluster centres (tissue types) for the given data samples (image intensity of pixels in the CT/MR scans) and probabilities (membership functions) for each data (image intensity of a pixel) belonging to the calculated cluster centres [22].

Once the tissue types (cluster centres) are determined, the material properties at the integration points of the computational grid can be calculated using the membership functions (probabilities) for each pixel (intensity at the integration point) and all tissue types (cluster centres) [19]. The computed material properties are 'fuzzy' rather than 'exact' values and there are no clear boundaries between tissues [20]. However, such 'fuzzy' material properties do not compromise the accuracy of prediction of tissue/organ deformations as it has been indicated in our previous studies that the material properties make a weak impact on the predicted deformations for problems where loading is prescribed as forced motion of the boundaries [14, 18].

A challenge in application of FCM for tissue classification is how to determine the tissue types in a given set of medical images. There is no standard criterion to determine how many tissue types (cluster centres) are needed to predict deformations with the accuracy sufficient (typically two-times the in-plane voxel size) for whole-body CT registration [20].

This chapter presents an application of the fuzzy tissue classification using the FCM algorithm for creating patient-specific whole-body biomechanical models for predicting tissue/organ deformations for registration of whole-body CT images and analyses sensitivity of the registration accuracy to the FCM algorithm parameters. For a whole-body CT image dataset, five major abdominal body organ/tissue types can be recognised from the CT images. As the lung is a large body organ and it can be easily distinguished from the CT images, we evaluated the accuracy of whole-body image registration using the patient-specific finite element model and the proposed fuzzy tissue classification method by comparing contours of the lung in the registered images (source images warped using deformations predicted by patient-specific biomechanical model) and target images, and the results show that misalignments are within two-times the image voxel size.

8.2 Methods

8.2.1 Whole-Body CT Image Dataset

The whole-body CT image dataset analysed in this study was acquired from the publicly available Slicer Registration Library (Case #20: Intra-subject whole-body/torso PET-CT http://www.na-mic.org/Wiki/index.php/Projects:RegistrationLibrary:RegLib_C20b). One image set is treated as moving/source image and another one is fixed/target image.

The original image dataset has resolution of 0.98 mm × 0.98 mm × 5 mm. Following our previous study [23], we resampled the image-sets to 1 mm × 1 mm × 2.5 mm using the built-in 'Resample Scalar Volume' tool in the 3D SLICER (http://www.slicer.org/)—an open-source software for visualisation, registration, segmentation and quantification of medical data developed by Artificial Intelligence Laboratory of Massachusetts Institute of Technology and Surgical Planning Laboratory at Brigham and Women's Hospital and Harvard Medical School [24].

8.2.2 Geometry Discretisation

The 3-D patient-specific geometry was extracted from CT images using the 'Intensity Threshold Segmentation' module available in the 3D SLICER software package [24]. Following [25], we used 8-noded hexahedral elements with one integration point to build the computational grid. Although tetrahedral meshes are popular [26], 8-noded hexahedral elements with one integration point do not exhibit volumetric locking that occurs when 4-noded tetrahedral elements are applied to incompressible/nearly incompressible continua such as soft tissues [27] and tend to offer better computational efficiency than tetrahedral meshes (for the same characteristic size of an element, less hexahedrons than tetrahedrons are needed to mesh a given volume). The computational grid consists of 51,479 elements and 55,944 nodes, as shown in Fig. 8.1.

8.2.3 Fuzzy C-Means Method for Tissue Classification

The FCM algorithm is used to calculate cluster centres for given data samples [22]. We have successfully used it in our previous studies to assign material properties at the integration points when computing brain deformations caused by craniotomy-induced brain shift [19], and for predicting the deformation field of body organs/tissues for whole-body CT image registration [20]. For tissue classification, the data samples are image intensity of all pixels in the image. Once the number of tissue types is selected, the FCM algorithm classifies pixels (data samples) as

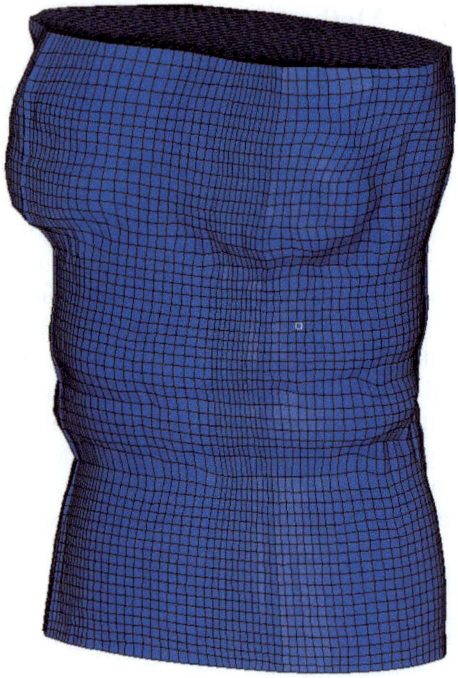

Fig. 8.1 Spatial discretisation of the whole-body geometry using hexahedral elements. The computational grid consists of 51,479 elements and 55,944 nodes. This grid was also used in our previous study [20]

belonging to different groups and calculates the cluster centre (image intensity) for each group by computing the membership functions (probabilities) that link image intensity at each pixel with all the specified tissue types, by minimising the objective function J_{FCM} [19, 22]:

$$J_{\mathrm{FCM}} = \sum_{i=1}^{N} \sum_{j=1}^{C} u_{ij}^{q} d\left(x_i, \theta_j\right) \tag{8.1}$$

where N is data samples (i.e. pixels in CT images), C is the number of cluster centres (tissue types/classes), q is the weighting factor referred to in the literature [28] as the fuzziness degree of clustering, μ_{ij} is the fuzzy membership function that expresses the probability of one data sample x_i (pixel) belonging to a specified cluster centre θ_j (tissue type/class) and d is the spatial distance between data sample x_i and cluster centre θ_j We used the fuzziness degree of clustering q of 2 which is a value commonly applied for soft tissue classification [29, 30].

The cluster centres θ_j (tissue classes) and the fuzzy membership functions μ_{ij} can be calculated by minimising the objective function (1). But, the number of cluster centres (tissue types/classes) C remains unknown. Therefore, we use a patient's whole-body CT image dataset as an example to analyse the selection of this parameter in detail in Sect. 8.3.

8.2.4 Loading and Boundary Conditions

In this study, computation of deformation field within the body organs/tissues is formulated as a displacement-zero-traction problem of computational mechanics where whole-loading is prescribed as forced motion of the boundary. For such formulation, the predicted deformations within the analysed continuum depend very weakly on the mechanical properties [11]. We select the spine (vertebrae) as the boundary to apply the forced motion as the spine can be easily distinguished from surrounding soft tissues in CT images. The displacement (forced motion) to align the spine in source and target images can be determined using rigid registration (we used 'Rigid Registration' function in the 3D SLICER [24]) for each vertebra.

We also considered using landmarks located on the skin as a source of information about motion of the boundary. However, as there are too few features on the abdominal skin to define such landmarks and they proved to be rather difficult to reliably locate/distinguish in the source and target images, no constraints and contact were applied to the body surface (skin) when conducting registration of the whole-body CT images. The proposed biomechanical model, however, allows for adding correspondence between easily distinguishable surface points as constraints if desirable.

8.2.5 Numerical Solutions

The non-linear patient-specific finite element model is solved using our previously developed Total Lagrangian Explicit Dynamics (TLED) finite element algorithm [25, 31, 32]. The algorithm utilises central difference method to discretise the temporal derivatives so that the discretised equations are integrated in stepping forward manner without any iteration. To accelerate the convergence to steady state, a dynamic relaxation is used [32]. For further improvement of computation efficiency, the TLED has been parallelised to harness computational power of Graphics Processing Units (GPUs) as shown in [33].

8.3 Results and Discussion

8.3.1 Image Intensity Distribution of Whole-Body CT Images

The FCM algorithm we applied to assign material properties at the integration points of the computational grid (finite element mesh) is a statistical feature classification method [34]. It calculates cluster centres (tissue types) and membership functions for data samples (i.e. image intensity of pixels depicted in CT images) using Eq. (8.1). As pointed out in Sect. 8.2.3, the number of cluster centres (tissue

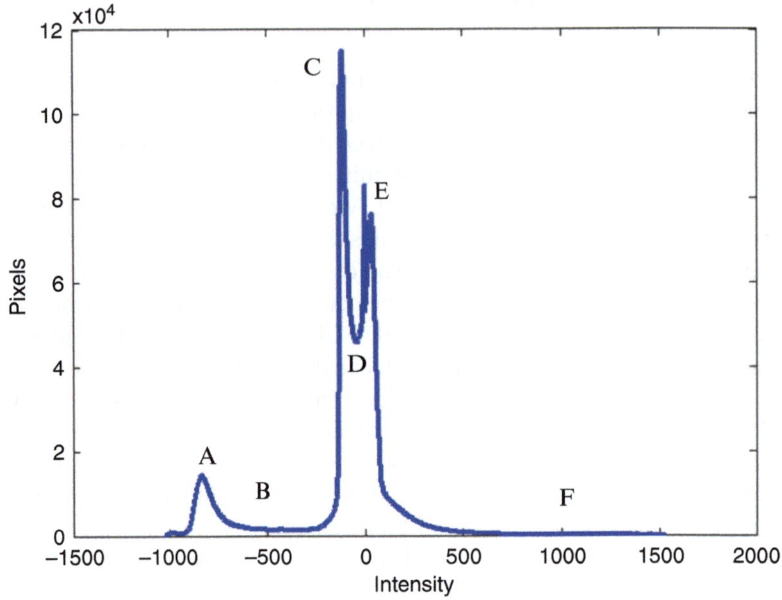

Fig. 8.2 Statistical distribution of image intensity for a patient's whole-body CT image dataset

Table 8.1 Tissue types and associated image intensity

Intensity	−900	−300	−80	−20	80	950
Statistical classification	Group A	Group B	Group C	Group D	Group E	Group F
Anatomical classification	Class 1: Gas-filled spaces (abdominal cavities)		Class 2: Fat	Class 3: Muscles (abdominal organs)	Class 4: Intestines (stomach)	Class 5: Bones

types/classes, parameter C in Eq. (8.1)) is an unknown parameter. Determining this number requires analysis of the statistical constituents of the CT images represented by distribution of the image intensity. Figure 8.2 shows the statistical distribution of image intensity for all pixels within a given patient's whole-body CT image dataset.

As can be seen in Fig. 8.2, in the image dataset analysed here, the intensity varies from −1200 to 1500 and divides into six groups: (1) most pixels are concentrated at four clusters (i.e. group A, C, D and E in Fig. 8.2, with the cluster centres at −850, −80, −20 and 80, respectively); (2) the remaining pixels are equally distributed between two groups (i.e. group B from −600 to −200 and group F from 400 to 1500). Therefore, a total of six statistical groups (clusters) can be distinguished and used for tissue classification (see Table 8.1).

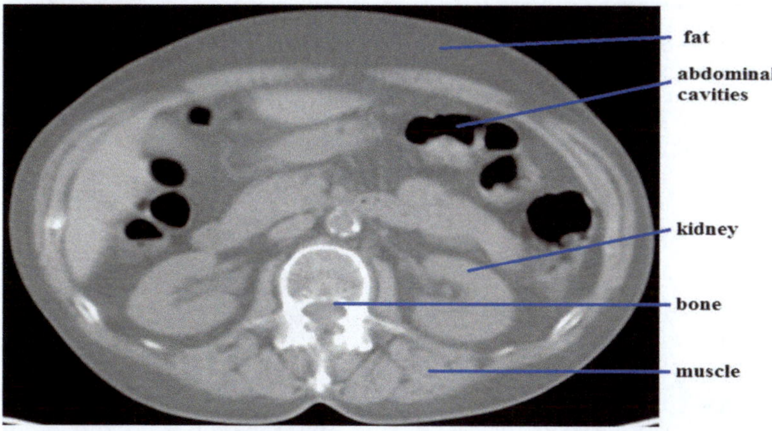

Fig. 8.3 A typical transverse section slice from a whole-body CT image dataset

Table 8.2 Material properties for tissue types

	Class 1	Class 2	Class 3	Class 4	Class 5
Shear modulus (kPa)	0.53	1.07	3.57	4.05	rigid
	[35]	[36]	[37–39]	[40]	

8.3.2 Tissue Classification

In Sect. 8.3.1, we analysed the statistical distribution of image intensity for a patient's whole-body CT image dataset. In this section, we relate this distribution to the anatomical constituents (i.e. major body organs/tissues) of the human body. Figure 8.3 shows a typical transverse slice of whole-body CT image. Average positions to different classes are: (1) gas-filled spaces (abdominal cavities); (2) fat; (3) muscles and abdominal organs (i.e. liver, kidneys); (4) intestines (stomach) and (5) bones. These five anatomical tissue types and their corresponding average image intensity are listed in Table 8.1. The shear modulus for these five anatomical tissue classes calculated using the FCM algorithm is given in Table 8.2.

The number of cluster centres (parameter C in Eq. (8.1)) is one of the parameters controlling the results of tissue classification in the FCM algorithm.

8.3.3 Parametric Study

Following Li et al. [20] and Mostayed et al. [41], we evaluate the registration accuracy by comparing the edges/contours of a body organ in registered (i.e. source image warped using the deformations predicted by the non-linear patient-specific finite element model) and target images. As the lung is a large body organ that can

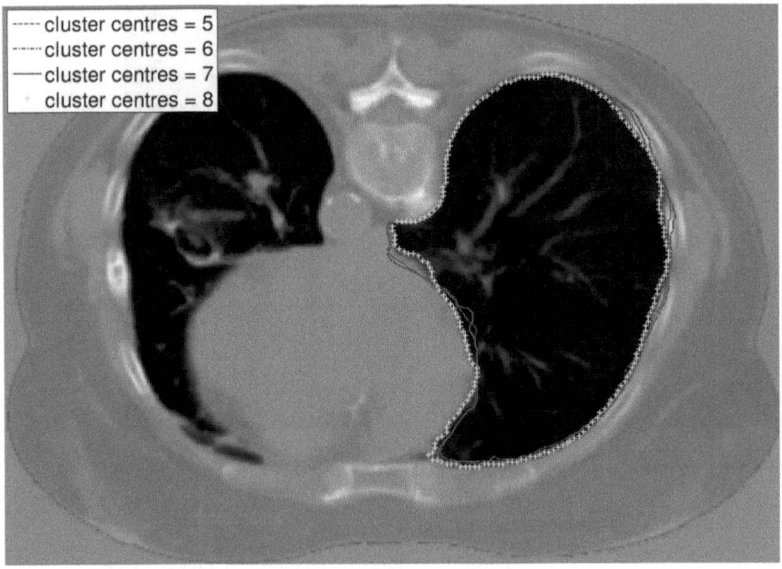

Fig. 8.4 Comparison of lung contours from the registered images (i.e. source image warped using deformations predicted by the non-linear patient-specific finite element model and the fuzzy tissue classification method). The *red dashed line* represents predicted deformations using 5 cluster centres; the *blue dashed line* represents predicted deformations using 6 cluster centres; the *pink solid line* represents predicted deformations using 7 cluster centres and the *blue star line* represents predicted deformations using 8 cluster centres

be reliably distinguished from the surrounding tissues, in this study we qualitatively evaluate the registration accuracy by comparing the contours of the lung extracted from the registered and target images. Following previous studies [15, 41], we consider any misalignment less than two-times the voxel size of the original source image to be successfully registered.

Sensitivity of the accuracy of whole-body image registration using non-linear patient-specific finite models to the number of cluster centres used by the FCM algorithm (parameter C in Eq. (8.1)) is demonstrated in Fig. 8.4. When the number of cluster centres is equal to eight (Table 8.3), the five major tissue classes can be successfully distinguished and the contour of the lung extracted from the registered images is very close to that extracted from the target image (i.e. distance between these two contours is within two-times the image voxel size, see Fig. 8.4). More cluster centres (i.e. 9, 10...) would increase the computational cost of the FCM algorithm without affecting/improving the results of tissue classification. For less than eight cluster centres the FCM algorithm is not able to distinguish five major tissue types, which results in somewhat poorer accuracy when predicting deformations of the body organs/tissues.

Table 8.3 Tissue classification with different cluster centres (tissue types)

Cluster centres	Image intensity							
5	−750		−400	−98		40	400	
6	−750	−600	−350	−96		40	400	
7	−750	−600	−350	−97		30	250	600
8	−750	−600	−350	−105	−30	40	250	600

8.4 Conclusions

Patient-specific biomechanical modelling for whole-body CT image registration often involves subjective and time-consuming image segmentation that divides whole-body CT scans into non-overlapping constituents with different material properties. To eliminate the need for tedious image segmentation, we have successfully developed a fuzzy tissue classification for creating the non-linear patient-specific biomechanical models without image segmentation. In this paper, we analysed the statistical and anatomical constituents for a patient's whole-body CT image dataset, and sensitivity of the registration accuracy for whole-body CT images using non-linear patient-specific finite element models to the FCM classification parameter.

The results suggest that when applying the FCM algorithm to assign material properties at the integration points of finite element mesh directly from the whole-body CT images. The number of cluster centres needs to be larger than the number of tissue types that needs to be distinguished. The accuracy of prediction of organ/tissue deformations when applying such models in whole-body CT image registration tends to be affected by the number of cluster centres and associated tissue classification. However, the effect is moderate and even for relatively small number of cluster centres prediction can be obtained.

To validate the accuracy of registration for whole-body CT images, in this study one patient's whole-body CT image dataset was analysed using the patient-specific non-linear finite element model and fuzzy tissue classification. Registration accuracy was qualitatively evaluated by comparing contours of the lung from the registered (i.e. source image warped using deformations predicted by non-linear patient-specific finite element model) images and target images. The misalignments were within two-times the image voxel size.

Acknowledgments The first author is a recipient of the APA scholarship and acknowledges the financial support of the University of Western Australia. The financial support of National Health and Medical Research Council (Grant No. APP1006031) and Australian Research Council (Discovery Grants No. DP1092893 and DP120100402) is gratefully acknowledged. In addition, the authors also gratefully acknowledge the support of National Centre for Image Guided Therapy (NIH U41RR019703). The whole-body CT image dataset analysed in this study was acquired from the publicly available Registration Library by the National Alliance for Medical Image Computing (NA-MIC). NA-MIC was funded by the National Institutes of Health through the NIBIB NIH HHS, Grant U54 EB005149.

References

1. A.V. D'Amico, R. Cormack, S. Kumar, C.M. Tempany, Real-time magnetic resonance imaging-guided brachytherapy in the treatment of selected patients with clinically localized prostate cancer. J. Endourol. **14**(4), 367–370 (2009)
2. S.K. Warfield, S.J. Haker, I.F. Talos, C.A. Kemper, N. Weisenfeld, A.U. Mewes, D. Goldberg-Zimring, K.H. Zou, C.F. Westin, W.M. Wells, C.M. Tempany, A. Golby, P.M. Black, F.A. Jolesz, R. Kikinis, Capturing intraoperative deformations: research experience at Brigham and Women's hospital. Med. Image Anal. **9**(2), 145–162 (2005)
3. M. Jenkinson, S. Smith, A global optimisation method for robust affine registration of brain images. Med. Image Anal. **5**(2), 143–156 (2001)
4. W.M. Wells 3rd, P. Viola, H. Atsumi, S. Nakajima, R. Kikinis, Multi-modal volume registration by maximization of mutual information. Med. Image Anal. **1**(1), 35–51 (1996)
5. A. Sotiras, C. Davatzikos, N. Paragios, Deformable medical image registration: a survey. IEEE Trans. Med. Imaging **32**(7), 1153–1190 (2013)
6. S. Oguro, K. Tuncali, H. Elhawary, P.R. Morrison, N. Hata, S.G. Silverman, Image registration of pre-procedural MRI and intra-procedural CT images to aid CT-guided percutaneous cryoablation of renal tumors. Int. J. Comput. Assist. Radiol. Surg. **6**(1), 111–117 (2011)
7. D. Rueckert, L.I. Sonoda, C. Hayes, D.L. Hill, M.O. Leach, D.J. Hawkes, Nonrigid registration using free-form deformations: application to breast MR images. IEEE Trans. Med. Imaging **18**(8), 712–721 (1999)
8. A. Hagemann, K. Rohr, H.S. Stiehl, U. Spetzger, J.M. Gilsbach, Biomechanical modeling of the human head for physically based, nonrigid image registration. IEEE Trans. Med. Imaging **18**(10), 875–884 (1999)
9. R.V. O'Toole III, B. Jaramaz, A.M. Digioia III, C.D. Visnic, R.H. Reid, Biomechanics for preoperative planning and surgical simulations in orthopedics. Comput. Biol. Med. **25**(2), 183–191 (1995)
10. J.G. Snedeker, S.H. Wirth, N. Espinosa, Biomechanics of the normal and arthritic ankle joint. Foot Ankle Clin. **17**(4), 517–528 (2012)
11. A. Wittek, G. Joldes, M. Couton, S.K. Warfield, K. Miller, Patient-specific non-linear finite element modelling for predicting soft organ deformation in real-time; application to non-rigid neuroimage registration. Prog. Biophys. Mol. Biol. **103**(2-3), 292–303 (2010)
12. G. Picinbono, H. Delingette, N. Ayache, Non-linear anisotropic elasticity for real-time surgery simulation. Graph. Model. **65**(5), 305–321 (2003)
13. J. Hu, X. Jin, J.B. Lee, L. Zhang, V. Chaudhary, M. Guthikonda, K.H. Yang, A.I. King, Intraoperative brain shift prediction using a 3D inhomogeneous patient-specific finite element model. J. Neurosurg. **106**(1), 164–169 (2007)
14. K. Miller, A. Wittek, G. Joldes, Biomechanical modeling of the brain for computer-assisted neurosurgery, in *Biomechanics of the Brain*, ed. by K. Miller (Springer, New York, 2011), pp. 111–136
15. R.R. Garlapati, A. Roy, G.R. Joldes, A. Wittek, A. Mostayed, B. Doyle, S.K. Warfield, R. Kikinis, N. Knuckey, S. Bunt, K. Miller, More accurate neuronavigation data provided by biomechanical modeling instead of rigid registration. J. Neurosurg. **120**(6), 1477–1483 (2014)
16. N. Sharma, L.M. Aggarwal, Automated medical image segmentation techniques. J. Med. Phys. **35**(1), 3–14 (2010)
17. L.K. Kwah, R.D. Herbert, L.A. Harvey, J. Diong, J.L. Clarke, J.H. Martin, E.C. Clarke, P.D. Hoang, L.E. Bilston, S.C. Gandevia, Passive mechanical properties of gastrocnemius muscles of people with ankle contracture after stroke. Arch. Phys. Med. Rehabil. **93**(7), 1185–1190 (2012)
18. K. Miller, J. Lu, On the prospect of patient-specific biomechanics without patient-specific properties of tissues. J. Mech. Behav. Biomed. Mater. **27**, 154–166 (2013)

19. J.Y. Zhang, G.R. Joldes, A. Wittek, K. Miller, Patient-specific computational biomechanics of the brain without segmentation and meshing. Int. J. Numer. Method Biomed. Eng. **29**(2), 293–308 (2013)
20. M. Li, A. Wittek, G. Joldes, G. Zhang, F. Dong, R. Kikinis, K. Miller, Whole-body image registration using patient-specific non-linear finite element model, in *Computational Biomechanics for Medicine: Fundamental Science and Patient-Specific Application*, ed. by B.J. Doyle, K. Miller, A. Wittek, P.M.F. Nielsen (Springer, New York, 2013), pp. 113–122
21. M. Li, K. Miller, G. Joldes, R. Kikinis, A. Wittek, Patient-specific meshless model for whole-body image registration, in *Biomedical Simulation*, ed. by F. Bello, S. Cotin, Volume 8789 of the series Lecture Notes in Computer Science (Springer International Publishing, 2014), pp. 50–57
22. J.C. Bezdek, R. Ehrlich, W. Full, FCM - the fuzzy C-means clustering-algorithm. Comput. Geosci. **10**(2–3), 191–203 (1984)
23. M. Li, K. Miller, G.R. Joldes, B. Doyle, R.R. Garlapati, R. Kikinis, A. Wittek, Patient-specific biomechanical model as whole-body CT image registration tool. Med. Image Anal. **22**(1), 22–34 (2015)
24. A. Fedorov, R. Beichel, J. Kalpathy-Cramer, J. Finet, J.C. Fillion-Robin, S. Pujol, C. Bauer, D. Jennings, F. Fennessy, M. Sonka, J. Buatti, S. Aylward, J.V. Miller, S. Pieper, R. Kikinis, 3D Slicer as an image computing platform for the Quantitative Imaging Network. Magn. Reson. Imaging **30**(9), 1323–1341 (2012)
25. K. Miller, G. Joldes, D. Lance, A. Wittek, Total Lagrangian explicit dynamics finite element algorithm for computing soft tissue deformation. Commun. Numer. Methods Eng. **23**(2), 121–134 (2007)
26. A. Wittek, K. Miller, R. Kikinis, S.K. Warfield, Patient-specific model of brain deformation: application to medical image registration. J. Biomech. **40**(4), 919–929 (2007)
27. S.C. Tadepalli, A. Erdemir, P.R. Cavanagh, Comparison of hexahedral and tetrahedral elements in finite element analysis of the foot and footwear. J. Biomech. **44**(12), 2337–2343 (2011)
28. M.A. Balafar, A.R. Ramli, M.I. Saripan, S. Mashohor, Review of brain MRI image segmentation methods. Artif. Intell. Rev. **33**(3), 261–274 (2010)
29. L.O. Hall, A.M. Bensaid, L.P. Clarke, R.P. Velthuizen, M.S. Silbiger, J.C. Bezdek, A comparison of neural network and fuzzy clustering-techniques in segmenting magnetic-resonance images of the brain. IEEE Trans. Neural Netw. **3**(5), 672–682 (1992)
30. D.L. Pham, J.L. Prince, Adaptive fuzzy segmentation of magnetic resonance images. IEEE Trans. Med. Imaging **18**(9), 737–752 (1999)
31. G.R. Joldes, A. Wittek, K. Miller, Suite of finite element algorithms for accurate computation of soft tissue deformation for surgical simulation. Med. Image Anal. **13**(6), 912–919 (2009)
32. G.R. Joldes, A. Wittek, K. Miller, Computation of intra-operative brain shift using dynamic relaxation. Comput. Methods Appl. Mech. Eng. **198**(41–44), 3313–3320 (2009)
33. G.R. Joldes, A. Wittek, K. Miller, Real-time nonlinear finite element computations on GPU - application to neurosurgical simulation. Comput. Methods Appl. Mech. Eng. **199**(49-52), 3305–3314 (2010)
34. S. Chuai-Aree, C. Lursinsap, P. Sophasathit, S. Siripant, Fuzzy C-mean: a statistical feature classification of text and image segmentation method. Int. J. Uncertainty Fuzziness Knowledge Based Syst. **9**(6), 661–671 (2001)
35. J. Alcaraz, L. Buscemi, M. Grabulosa, X. Trepat, B. Fabry, R. Farre, D. Navajas, Microrheology of human lung epithelial cells measured by atomic force microscopy. Biophys. J. **84**(3), 2071–2079 (2003)
36. A. Samani, J. Zubovits, D. Plewes, Elastic moduli of normal and pathological human breast tissues: an inversion-technique-based investigation of 169 samples. Phys. Med. Biol. **52**(6), 1565–1576 (2007)
37. K. Miller, Constitutive modelling of abdominal organs. J. Biomech. **33**(3), 367–373 (2000)
38. A.M. Collinsworth, S. Zhang, W.E. Kraus, G.A. Truskey, Apparent elastic modulus and hysteresis of skeletal muscle cells throughout differentiation. Am. J. Physiol. Cell Physiol. **283**(4), C1219–C1227 (2002)

39. J. Rosen, J.D. Brown, S. De, M. Sinanan, B, Hannaford biomechanical properties of abdominal organs in vivo and postmortem under compression loads. J. Biomech. Eng. Trans. ASME **130**(2), 021020-1-17
40. Y.J. Lim, D. Deo, T.P. Singh, D.B. Jones, and S. De, In situ measurement and modeling of biomechanical response of human cadaveric soft tissues for physics-based surgical simulation, Surg Endosc, **23**(6): 1298–1307 (2009)
41. A. Mostayed, R.R. Garlapati, G.R. Joldes, A. Wittek, A. Roy, R. Kikinis, S.K. Warfield, K. Miller, Biomechanical model as a registration tool for image-guided neurosurgery: evaluation against BSpline registration. Ann. Biomed. Eng. **41**(11), 2409–2425 (2013)

Chapter 9
GPU-Based Fast Finite Element Solution for Nonlinear Anisotropic Material Behavior and Comparison of Integration Strategies

Vukašin Štrbac, David M. Pierce, Jos Vander Sloten, and Nele Famaey

9.1 Introduction

Finite element (FE) simulations are increasingly employed to assess and improve the performance of biomedical devices and procedures. Examples are performance analyses of stents [1, 2] or arterial clamping device design optimization [3]. Speed requirements are less stringent when these analyses are performed pre- or postoperatively. Even so, for models with noteworthy geometrical and material complexity, solution speed is an important barrier keeping these simulations from being integrated into the clinical workflow. Moreover, also intraoperative utilization of finite element analysis is increasingly proposed as a solution, i.e. for soft tissue overload prevention [4] or for brain shift estimation [5]. For these situations, but also for virtual surgical simulators, a continuum-mechanical representation has long been impossible due to the required update rates. In the past decade, alongside the evolution of computational architectures and advancement of computational power, a number of fast FE solvers have appeared in the literature [6–8]. To the best of the authors' knowledge, Miller et al. [9] were the first to publish research on an efficient nonlinear real-time FE solution to intraoperatively provide information about tissue response on patient-specific anatomy during actual surgery. The current paper expands upon the Total Lagrangian Explicit Dynamic (TLED) algorithm proposed

V. Štrbac (✉) • J.V. Sloten • N. Famaey
Biomechanics Section, KU Leuven, Heverlee, Belgium
e-mail: vule.strbac@kuleuven.be; Jos.Vandersloten@kuleuven.be; nele.famaey@kuleuven.be
https://www.mech.kuleuven.be/en/bme/research/soft-tissue-biomechanics/home

D.M. Pierce
Department of Mechanical Engineering, University of Connecticut, Storrs, CT, USA
e-mail: dmpierce@engr.uconn.edu
http://im.engr.uconn.edu

G. Joldes et al. (eds.), *Computational Biomechanics for Medicine*,
DOI 10.1007/978-3-319-28329-6_9

by this research group, presenting results of an efficient 3D implementation on the GPU using Nvidia CUDA technology. In particular, the accuracy and solution speed of the simulation of an expansion of an artery modeled with a nonlinear anisotropic material using our implementation are compared to that of an established implicit FE solver FEAP (University of California, Berkeley, USA). Furthermore, differences in computation speed between different element integration types and between isotropic or anisotropic material models are evaluated.

9.2 Materials and Methods

9.2.1 3D TLED Algorithm on the GPU

TLED is an explicit dynamic, large strain solver that uses the initial configuration of the domain as the reference configuration. The work-conjugate Second Piola-Kirchhoff stress and Green strain are used, and the primary kinematic variables are the deformation gradient and displacements. This combination is suitable for hyperelastic materials which are generally given in total rather than rate form. An additional advantage of the algorithm is that shape function derivatives with respect to the initial configuration are constant and can be computed before the time-marching phase. For a detailed description of TLED, see [9, 10].

The conventional implementation of TLED on GPUs is split into several kernels (parallel functions running on the GPU hardware) enabling the processing of large numbers of elements concurrently. This paradigm is used in all phases of the solver algorithm: computation of internal forces, time-marching, and the imposition of boundary conditions. Similar in function and benefits to distributed computing systems used for large simulations, CUDA parallelization is light-weight, tightly coupled and has much higher granularity in the work units and threads executing together. The management of GPU internals are left largely to the programmer. These involve single and double precision computing units, different memory types of different bandwidths, latencies, and capacity, instruction issue pipelines and schedulers, etc. For a more detailed description of CUDA TLED implementation, consult, e.g., [11, 12] and for in-breadth testing of TLED on GPUs, see [4].

9.2.2 Element Implementations

Three element types were implemented into the current platform, all of which are of the tri-linear hexahedral type, with varying integration procedures.

Under-Integrated Linear Hexahedron Classically used in TLED, the under-integrated (UI) linear hexahedron uses first order integration of the stress tensor at the centroid of the element. The combination of the single integration point and the first order isoparametric formulation leads to some strain modes of the element being stressless—a phenomenon called hourglassing. Hourglassing can be remedied to an extent, however, by applying one of several additional algorithms: basic orthogonal hourglass control [13] and its shift to a total Lagrangian worldview [14], or the more advanced assumed strain stabilization [15] used in several commercial tools. Suboptimal behavior for complex anisotropic materials can be expected, which can be remedied by increasing the mesh density used to discretize the boundary value problem. Under-integration is the cheapest solution in terms of both memory utilization and mathematical operations required, even when accounting for basic hourglassing schemes. This element is used as a benchmark reference for computational requirements and speed.

Fully Integrated Linear Hexahedron Full integration (FI) was also implemented into the TLED code, using the conventional second order Gaussian quadrature formula. Being the quintessential integration scheme for the tri-linear hexahedron, it proves very practical for validation and comparison purposes to other solvers. A great benefit of performing the integration in the element's natural coordinate space is the fact that in this configuration the shape function derivatives at points sampled by the full integration method are a simple function of the shape function derivatives at the centroid, and can easily be computed on-the-fly rather than being brought in by expensive memory fetching operations. Despite the accuracy provided by this element formulation, incompressibility or near-incompressibility of this element formulation often leads to pressure locking, increasing the volumetric stiffness unnaturally, and significantly deteriorating the quality of the results. From a computational standpoint, relative to the UI linear hexahedron, this element is eight times more expensive (conservative) in terms of operations and carries a minimum of twice the expense in memory. The latter point depends heavily on the implementation.

Selective-Reduced Linear Hexahedron The third element formulation that was implemented is the selective-reduced integration (SRI) element. Here, a full integration of the deviatoric terms of the stress tensor is performed in the eight integration points of the element and, additionally, a first order integration is done at the centroid for volumetric terms. The two phases are integrated separately and added to the total internal force vector. This element does not experience pressure locking, but is susceptible to shear-locking. As with the fully integrated element, it benefits from the same on-the-fly calculation of the shape function derivatives at the points sampled for the deviatoric terms, while central values are ready for use. Nevertheless, it is the most demanding element type to compute, requiring approximately nine times more operations than the under-integrated element and slightly higher memory requirements than the fully integrated element.

For all elements, per the mathematical theory of Gaussian integration, the actual integration is performed in the element's natural coordinate space. A pull back of the initial configuration (used in total Lagrangian) is necessary by way of the element Jacobian, previously computed for the deformation gradient.

Pressure Quadrilateral Pressure loading has also been added to the base TLED algorithm. Quadrilaterals belonging to the loaded surface are processed and final discrete force contributions are computed using full integration over elements' surfaces. This deformation-dependent distributed load is the only segment of the solver computed using updated Lagrangian values.

9.2.3 Material Implementations

Two nonlinear hyperelastic material models are implemented, both of which are described by their strain energy density function (SEDF) Ψ. This function can be additively decomposed into a deviatoric and a volumetric part as

$$\Psi = \Psi_{\text{dev}} + \Psi_{\text{vol}}. \tag{9.1}$$

For both material models, the volumetric component of the SEDF was defined as

$$\Psi_{\text{vol}} = \frac{K}{2}(J-1)^2, \tag{9.2}$$

with K the bulk modulus and J the determinant of the deformation gradient.

The second Piola-Kirchhoff stress tensor $\mathbf{S}$ can be derived from the SEDF as

$$\mathbf{S} = 2\frac{\partial\Psi}{\partial\mathbf{C}}. \tag{9.3}$$

neo-Hookean Material The first material model implemented is the isotropic hyperelastic neo-Hookean model $\Psi_{\text{dev}} = \Psi_{\text{dev}}^{\text{iso}}$, using the following form for the deviatoric part of the SEDF:

$$\Psi_{\text{dev}}^{\text{iso}} = \frac{\mu}{2}(\bar{I}_1 - 3), \tag{9.4}$$

where μ is the shear modulus and $\bar{I}_1$ is the first invariant of the deviatoric part of the right Cauchy-Green deformation tensor $\overline{\mathbf{C}}$.

GHO Model The second material model implemented is the Gasser–Holzapfel–Ogden (GHO) model, which describes a fiber-reinforced anisotropic material that also accounts for dispersion of the fibers. The deviatoric component of the SEDF contains an isotropic and an anisotropic term, $\Psi_{\text{dev}} = \Psi_{\text{dev}}^{\text{iso}} + \Psi_{\text{dev}}^{\text{ani}}$, corresponding to the matrix material and to the collagen fiber families, respectively [16]. The isotropic term corresponds to the deviatoric term of the neo-Hookean material [Eq. (9.4)] while the anisotropic term corresponds to

$$\Psi_{\text{dev}}^{\text{ani}} = \sum_{i=4,6} \frac{k_1}{2k_2}\left[e^{k_2(\kappa\bar{I}_1+(1-3\kappa)\bar{I}_i-1)^2} - 1\right], \tag{9.5}$$

where $k_1 > 0$ is a stress-like parameter, $k_2 > 0$ is a dimensionless parameter, and $\kappa \in \left[0, \frac{1}{3}\right]$ is a parameter related to the dispersion of the fibers, with the lower

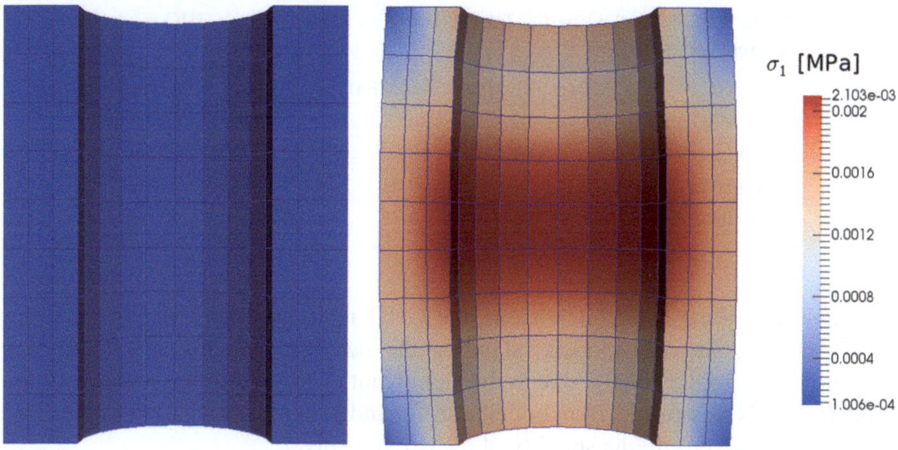

Fig. 9.1 Initial and final geometry of a test cylinder, using locked (top and bottom planes fixed in all degrees of freedom) boundary conditions, under-integrated elements, and a neo-Hookean material model

and upper limit corresponding to no fiber dispersion and fully dispersed fibers (i.e., isotropy), respectively. The material parameter, angle (ϕ), is defined between the local circumferential direction and the two fiber directions, symmetrically, in the plane of the material.

9.2.4 Sample Problem

A cylindrical mesh consisting of 540 elements (three layers of 180 elements) and 800 nodes was created to test the performance of different material implementations and elements w.r.t. accuracy and their relative speed of execution (Fig. 9.1). The total height of the cylinder is 90 mm, inner and outer diameters are 40 mm and 70 mm, respectively. The simulation is driven by a pressure boundary condition on the inside surface of the mesh, while the top and bottom kinematic boundary conditions have two configurations: "free," where nodes on the top and bottom planes (the extremes in z-direction) are constrained in the z-direction but are otherwise free to displace, and "locked," where the top and bottom planes are fixed in all degrees of freedom. The free boundary condition additionally has select nodes locked in the x- or y-direction to prevent rigid body rotation, but does not disturb the expected symmetry in the solution. The model with the locked boundary conditions accentuates shear stresses in the results, as shear is largely absent in the free expansion configuration.

Materials used are the isotropic neo-Hookean and the anisotropic GHO fiber-reinforced arterial tissue model. The material properties used can be found in Table 9.1. Note particularly the relation of shear and bulk moduli, always yielding a Poisson's ratio of 0.4995 used to enforce near-incompressibility.

Table 9.1 Material properties for neo-Hookean model, and the GHO model in three discrete layers

Material	μ (MPa)	K (MPa)	k_1 (MPa)	k_2	κ	ϕ(deg)
neo-Hookean	0.010	10	–	–	–	–
GHO layer 1	0.044	44	10.1	0.0	0.25	40.5
GHO layer 2	0.028	28	0.81	12.4	0.18	39.1
GHO layer 3	0.010	10	0.38	3.35	0.11	40.6

Pressure loading for models using the GHO material model is performed up to 120 mmHg, whereas models including the neo-Hookean material model are pressurized up to 8.5 mmHg, due to the significant difference in the stiffness of the models. In this way results from both the stiffer and softer materials lead to similarly large strains and ensure the stability of the explicit solver.

The solver uses a simple constant-step central-differences formula for time-marching, as per initial work on TLED [9]. The standard termination criterion for explicit dynamic simulations, based on the balance between internal, external, and inertial work was circumvented here. A simpler criterion, based on the root-mean-square error to an already known solution has been utilized. This method, however, still necessitates a manual check as the current solution might pass through the correct configuration during the solution phase, especially given the dynamic nature of the solver and the selected sample problem. All runs present in this study are performed in double precision floating point accuracy and use a time step of 5E−6 s. Load is imposed using a smooth loading curve in the total time duration of 0.01 s. Damping is implemented as per [4, 17] with a damping parameter (convergence rate) of 0.999.

In a first phase, results were obtained for purposes of testing the accuracy of our implementation. To this end, the aforementioned boundary value problems were also run in the implicit solver FEAP v8.2.k, using fully integrated elements, such that the final displacements could be compared.

The PC hardware used in this research contains an i7-4790K @4 GHz, 16 Gb of RAM, and an Nvidia GTX980 GPU, built upon the Maxwell architecture. Both FEAP and the CUDA code are run on the same machine.

9.3 Results

Accuracy is measured by computing the absolute (RMS) and relative root-mean-square (RRMS) values between the ground-truth FEAP solutions and our solver. As previously mentioned, the termination of CUDA simulations at a certain step was determined manually, at the point where the RMS error is constant. Accuracy, timings, and termination steps are shown in Table 9.2.

Comparative speed tests were performed on the set of CUDA results only. Due to large amount of data output for post-processing, timing was separated into pure

Table 9.2 Accuracy verification for the neo-Hookean (n-H, run for 1.5E4 steps) and GHO (run for 1.5E4 steps) material models using full integration

Simulation	RMS (mm)	RRMS	Wall-clock (s)	GPU time (s)	FEAP Wall-clock (s)
n-H locked	3.19E−6	1.14E−5	17.3	12.6	102.73
n-H free	3.15E−6	1.00E−5	20.4	12.6	108.50
GHO locked	4.68E−6	2.36E−5	14.2	11.2	92.68
GHO free	5.51E−6	2.72E−5	13.5	10.3	103.03

Root mean square and relative root mean square for accuracy verification, timings for general comparison

Table 9.3 Execution speed comparison between different integration routines, relative to the fastest solution, that of the under-integrated element with the neo-Hookean material model

Simulation	Wall-clock (s)	GPU time (s)	Speed
neo-Hooke UI	7.28	2.85	1.000
neo-Hooke FI	20.40	12.60	4.421
neo-Hooke SRI	18.16	13.44	4.715
GHO UI	7.59	3.15	1.105
GHO FI	20.08	15.32	5.375
GHO SRI	20.04	16.07	5.638

computational time performed by the solver and the total wall-clock time. Wall-clock includes both the solver and input/output (I/O) operations (of user-controlled frequency) that include DRAM-RAM and RAM-HDD (Hard Disk Drive) communication for post-processing. The solver run-time (GPU time) therefore includes the aggregate durations of kernels stated in Sect. 9.2.1, including all communication and arithmetic operations pertaining only to GPU internals and excludes all I/O operations. In-effect, GPU time measures only the solver time, since the entirety of the core solver is ported to the GPU. This time measurement was performed by an event-driven high-resolution clock exposed to the user by the CUDA framework, and provided by the GPU hardware (*cudaEventElapsedTime(...)*). All results in terms of speed are normalized to the computation time of the fastest solution—the under-integrated neo-Hookean (n-H)—and run for 15,000 steps. Speed testing results are shown in Table 9.3.

9.4 Discussion and Future Work

The accuracy results presented in Table 9.2 demonstrate the correctness of the implementation. The measured accuracy is at its realistic limit, given that FEAP outputs five significant digits by default. Note that these measurements were performed only between fully integrated elements in the two solvers since FEAP does not have an under-integrated or SRI element. Comparison between elements of different integration schemes would not result in accurate comparison of solvers, rather in the testing of element formulations.

The only difference between the presented material models is in the response of the two fiber families, i.e. in the anisotropy of the GHO material model. In comparison to the base matrix material (serving as total material response in neo-Hookean), the fiber response calculations are numerous. Additional memory transactions and storage required for intermediate values associated with the response of the fibers are substantial, exceeding those for neo-Hookean and suggesting that execution speeds should be significantly slower. Interestingly, tests for FI and SRI between the two materials show only a 10–21 % increase relative to UI, a result reflecting the complex, redundant, and autonomous execution and memory management of the CUDA GPU. It also reflects the fact that, when running in double precision, even the neo-Hookean material and the most basic linear hexahedral element exceed the available per-thread register memory of the device, resulting in register spilling and heavy use of caching. For comparison, a single precision solution for the UI neo-Hookean is approximately 1/4 (on the GTX980 [4]) that of the double precision solution time for a mesh of similar size, and is the fastest solution if not for accuracy considerations. Note that the presented mesh is not of sufficient density to induce maximal occupancy of the device, as it was not considered pertinent to the current work's scope; larger meshes would invariably produce higher GPU-CPU speedups. The above reference also shows results on a range of meshes, in conditions of under- or full-utilization of the device.

Materials approaching incompressibility, nearly ubiquitous in soft tissue, have a significant and adverse effect on the reduction of the critical time step used in simulations solved with explicit schemes. The GHO model is particularly sensitive to the ratio of bulk to shear modulus, i.e. the Poisson's ratio, as it has to be set high to keep the deformation nearly isochoric, particularly important in anisotropic material models. This sensitivity is much higher in such models than for the isotropic neo-Hookean [18, 19]. Future work should be aimed at an implementation of the mixed-element formulations using assumed strain stabilization and an improved termination criterion based on energy balance.

Acknowledgements This research was funded by an FP7 STREP project, CASCADE, by an interdisciplinary research project of KU Leuven (IDO), and by the Research Foundation Flanders (FWO).

References

1. F. Migliavacca, L. Petrini, M. Colombo, F. Auricchio, R. Pietrabissa, Mechanical behavior of coronary stents investigated through the finite element method. J. Biomech. **35**(6), 803–811 (2002) [Online]. Available: http://www.sciencedirect.com/science/article/pii/S0021929002000337
2. C. Conway, F. Sharif, J. McGarry, P. McHugh, A computational test-bed to assess coronary stent implantation mechanics using a population-specific approach. Cardiovasc. Eng. Technol. **3**(4), 374–387 (2012) [Online]. Available: http://dx.doi.org/10.1007/s13239-012-0104-8

3. N. Famaey, G. Sommer, J.V. Sloten, G.A. Holzapfel, Arterial clamping: finite element simulation and in vivo validation. J. Mech. Behav. Biomed. Mater. **12**, 107–118 (2012) [Online]. Available: http://www.sciencedirect.com/science/article/pii/S1751616112000902
4. V. Strbac, J. Vander Sloten, N. Famaey, Analyzing the potential of GPGPUs for real-time explicit finite element analysis of soft tissue deformation using CUDA. Finite Elem. Anal. Des. **105**, 79–89 (2015)
5. G.R. Joldes, A. Wittek, M. Couton, S.K. Warfield, K. Miller, Real-time prediction of brain shift using nonlinear finite element algorithms. Med. Image Comput. Comput. Assist. Interv. **12**(2), 300–307 (2009)
6. E.D. Cris Cecka, Adrian J. Lew, Assembly of finite element methods on graphics processors. Int. J. Numer. Methods Eng. **85**, 640–669 (2011)
7. Z.A. Taylor, M. Cheng, S. Ourselin, High-speed nonlinear finite element analysis for surgical simulation using graphics processing units. IEEE Trans. Med. Imaging **27**(5), 650–663 (2008) [Online]. Available: http://ieeexplore.ieee.org/stamp/stamp.jsp?arnumber=4388142
8. S.F. Johnsen, Z.A. Taylor, M.J. Clarkson, J. Hipwell, M. Modat, B. Eiben, L. Han, Y. Hu, T. Mertzanidou, D.J. Hawkes, S. Ourselin, Niftysim: a GPU-based nonlinear finite element package for simulation of soft tissue biomechanics. Int. J. Comput. Assist. Radiol. Surg. **10**(7), 1077–1095 (2014)
9. K. Miller, G. Joldes, D. Lance, A. Wittek, Total Lagrangian explicit dynamics finite element algorithm for computing soft tissue deformation. Commun. Numer. Methods Eng. **23**(2), 121–134 (2007)
10. T. Belytschko, W.K. Liu, B. Moran, *Nonlinear Finite Elements for Continua and Structures* (Wiley, New York, 2000)
11. G.R. Joldes, A. Wittek, K. Miller, Real-time nonlinear finite element computations on GPU - application to neurosurgical simulation. Comput. Methods Appl. Mech. Eng. **199**(49–52), 3305–3314 (2010) [Online]. Available: http://dx.doi.org/10.1016/j.cma.2010.06.037
12. G. R. Joldes, A. Wittek, K. Miller, Real-time nonlinear finite element computations on GPU: handling of different element types, In *Computational Biomechanics for Medicine*, ed by A. Wittek, P.M.F. Nielsen, K. Miller (Springer, New York, 2011), pp. 73–80
13. D.P. Flanagan, T. Belytschko, A uniform strain hexahedron and quadrilateral with orthogonal hourglass control. Int. J. Numer. Methods Eng. **17**, 679–706 (1981)
14. G.R. Joldes, A. Wittek, K. Miller, An efficient hourglass control implementation for the uniform strain hexahedron using the total Lagrangian formulation. Commun. Numer. Methods Eng. **24**, 1315–1323 (2008)
15. T. Belytschko, L.P. Bindeman, Assumed strain stabilization of the eight node hexahedral element. Comput. Methods Appl. Mech. Eng. **105**, 225–260 (1993)
16. T.C. Gasser, R.W. Ogden, G.A. Holzapfel, Hyperelastic modelling of arterial layers with distributed collagen fibre orientations. J. R. Soc. Interface **3**(6), 15–35 (2006)
17. G.R. Joldes, A. Wittek, K. Miller, An adaptive dynamic relaxation method for solving nonlinear finite element problems. application to brain shift estimation. Int. J. Numer. Methods Biomed. Eng. **27**(2), 173–185 (2011) [Online]. Available: http://dx.doi.org/10.1002/cnm.1407
18. C. Sansour, On the physical assumptions underlying the volumetric-isochoric split and the case of anisotropy. Eur. J. Mech. A Solids **27**, 28–39 (2008)
19. J. Helfenstein, M. Jabareen, E. Mazza, S. Govindjee, On non-physical response in models for fiber-reinforced hyperelastic materials. Int. J. Solids Stuct. **47**, 2056–2061 (2010)

Chapter 10
Fast Prediction of Femoral Biomechanics Using Supervised Machine Learning and Statistical Shape Modeling

Elham Taghizadeh, Michael Kistler, Philippe Büchler, and Mauricio Reyes

10.1 Introduction

Osteoporosis is a very frequent disease that affects the life of many people after the age of 50. Osteoporosis causes annually more than 2.3 million fractures in Europe and in the USA. In 2002, it was reported that in England and Wales, the osteoporosis related fractures cost £942 million annually and this value would increase with the ageing of the population in the western countries [1]. An accurate estimation of bone strength and fracture risk can help the diagnosis of osteoporosis, leading to an improvement of patient's quality of life and reduced associated healthcare costs.

Dual energy X-ray absorptiometry (DEXA) scan is the standard clinical diagnostics tool to evaluate the level of osteoporosis and the related risk of fracture. A T-score smaller than or equal to -2.5 of femoral neck or lumbar spine indicates osteoporosis. T-score is the number of standard deviations (STD) that bone mineral density (BMD) deviates from the average of BMD, measured in a healthy 30-year-old population with the same gender and ethnicity as the patient [1].

To automate the diagnosis of osteoporosis from DEXA images, Whitmarsh and colleagues used statistical shape and appearance models. They proposed a Fisher Linear Discriminant Analysis (FLDA) method to classify bones having a high or low fracture risk [2]. Sarkalkan and colleagues proposed 2D finite element models built from DEXA images to predict the fracture risk of the proximal femur [3].

It has been shown that 3D (FE) analyses predict bone strength more accurately than clinical methods such as DEXA [4]. However, the adoption of FE analyses into clinical practice has been hampered by its computational complexity and required technical competences. To analyze the bone behavior under a certain loading

---

E. Taghizadeh (✉) • M. Kistler • P. Büchler • M. Reyes
Institute for Surgical Technology and Biomechanics, University of Bern, Bern, Switzerland
e-mail: elham.taghizadeh@istb.unibe.ch; michael.kistler@istb.unibe.ch; philippe.buechler@istb.unibe.ch; mauricio.reyes@istb.unibe.ch

G. Joldes et al. (eds.), *Computational Biomechanics for Medicine*,
DOI 10.1007/978-3-319-28329-6_10

condition, an accurate segmentation of the bone is necessary, a valid finite element mesh must be generated, and appropriate boundary conditions need to be applied to the model. These preparation steps are followed by time-consuming calculations to determine the biomechanical behavior of the bone. All of these steps are time consuming and computationally demanding, which make the FE analysis less appealing for clinicians [5]. As a consequence, up to now, the FE analysis techniques did not reach the clinical workflow. Different research studies aimed to automate the segmentation [6] and finite element mesh creation [7, 8], however to the best of our knowledge no method has been proposed to bypass the computational complexity of FE calculations. In this paper we aim at alleviating the aforementioned issues of FE analysis to promote their adoption into clinical practice.

The two most important features describing bone biomechanics are shape and BMD. Therefore, we hypothesize that machine learning techniques can be used to predict the biomechanical properties of the bone using shape and density features extracted from clinical patient scans, as well as patient anthropometric information. To this end, we propose a supervised learning approach to predict the outcome of FE analysis. As feature predictors for bone shape and density, we propose to characterize this information in a compact way by using statistical shape modeling of the anatomy [6]. In this way, we reduce the dimensionality of the feature space, which leverages the building process of the machine learning model, and moreover, allows us to exploit previous developments in statistical shape modeling (e.g., active shape models [9]). We demonstrate this by predicting bone stresses from clinical CT (*FEP*), where features are extracted from a statistical shape and statistical intensity model of the human femur and patient's anthropometric information. As a second demonstration we present preliminary results on a simplified scenario where FE femur biomechanics are predicted from 2D X-ray images. Morphometric and density information available in the 2D image was used as predictors.

In the next section we describe in detail how the prediction models are built, how features are defined and extracted, and one example scenario to demonstrate how the approach can be adapted for X-ray scans. In the Results section, the databases used for training and testing of the method are presented and the quality of the prediction is quantified. We conclude the paper by discussing the advantages and limitations of our proposed approach.

10.2 Method

In this section we explain the proposed method for finite element prediction, termed here *FEP*. We then follow by exemplifying how the proposed method can be employed for a different image modality, such as X-ray.

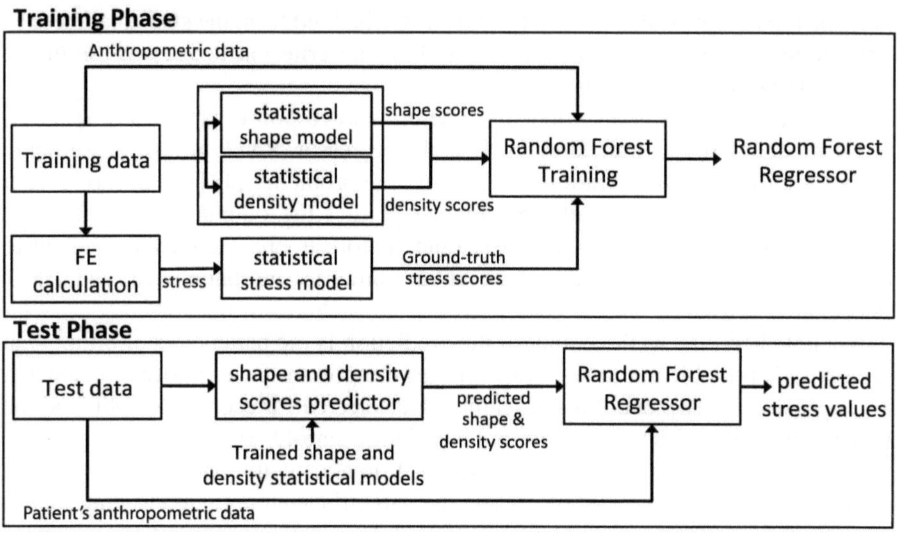

Fig. 10.1 Schematic description of the *FEP* predictions. Using shape, density, and stress scores of training data, we learn a random forest regression model. In the test phase, the trained random forest predicts from anthropometric and SSM-based bone and density predictors, extracted for a new image, the parameters of the statistical stress model

10.2.1 Finite Element Prediction Framework

The main framework for *FEP* is summarized in Fig. 10.1. Following the same scheme as in supervised learning, our approach has two stages.

First, during the training stage, a statistical model of shape and intensity is created as in [10]. In short, an iterative mesh morphing method [11] is used to compute point correspondences for a dense volumetric mesh consisting of approximately 190,000 nodes and 130,000 tetrahedral elements. Bone density for each node is then extracted from the original CT scans [10]. A principal component analysis (PCA) is then performed separately on shape and density information, yielding two separate models. As shown in Fig. 10.1 (training phase) each bone can then be modeled through shape and density scores.

As response variables, FE computations are used to calculate stress values on each node of the FE mesh. The FE analyses were performed with the commercial package Abaqus/Standard (Abaqus v6.12, SIMULIA, USA). Boundary conditions (BC) representing a walking situation were applied to the bone models. We chose the loadings of the joint configuration proposed in [12], where the node constraints are selected at the femoral head, the intercondylar femoral notch, and the lateral epicondyle of the femur. The force values were calculated based on the body weight. The calculated nodal stress values were used to build a statistical model of stress.

A statistical model of the stress in the model was built. The scores of this model were used as output of the prediction algorithm. For the calculations, we considered only the top modes of shape, density, and stress obtained from the statistical models. The number of modes included was based on the criterion to keep 98 % of the variation that was in the dataset.

Using the set of aforementioned predictors and response variables, a random-forest model [13] was trained to work as the regressor. Random forests are being used for different classification and regression problems [14]. They are robust to noise and more importantly are able to predict the output even when some input information is missing. Besides, random forests are naturally conceived to use nature of different data, as here anthropometric, morphometric, and BMD information is used.

We note here that as the output of the prediction is the parameters of orthogonal vectors, it is possible to train one random forest regressor for each stress parameter. As suggested in [13], one-third of features are selected for each node-split, and the maximum depth for the tree is selected based on a tenfold cross validation.

During the test phase, given a patient CT image of the anatomy, the feature extraction process consists of projecting the patient's anatomy into the shape space to recover shape and density parameters [6, 10]. Here is where current and advanced SSM-based technologies (e.g., active shape models, hierarchical shape models [6]) can be used to compute scores for shape and density information. For the sake of simplicity we relied our experiments on a leave-one-out (LOO) scheme where these parameters are extracted during model building. We also included anthropometric features such as patient's age, gender, height, and weight in the input features. Finally, after feature extraction, FE predictions can be computed to yield stress scores, which are converted into stress values by simply drawing the corresponding sample values from the statistical model of stresses.

10.2.2 FEP for X-ray Images

By employing statistical shape and density scores to represent the anatomy and predict bone biomechanics, it is possible to decouple the prediction model from the input image modality. In other words, bone shape and density scores act as a "bridge" connecting the image modalities used to capture bone shape and density information of the patient to the image modality (CT scan) used to characterize bone biomechanics. As an example of using *FEP* for a different image modality, we demonstrate in this paper the case of having X-ray images as the input image modality used to capture bone shape and density information. We then show how to connect this information to shape and density scores used by *FEP* to predict bone biomechanics.

For the sake of simplicity, in this study we built synthetic X-ray scans by projecting the captured CT scans to two orthogonal planes. To characterize bone shape and density information, we used a set of simple yet effective feature descriptors. From two orthogonal X-ray images a total of 21 bone morphometric dimensions, as shown in Fig. 10.2, are extracted by selecting a few landmarks from

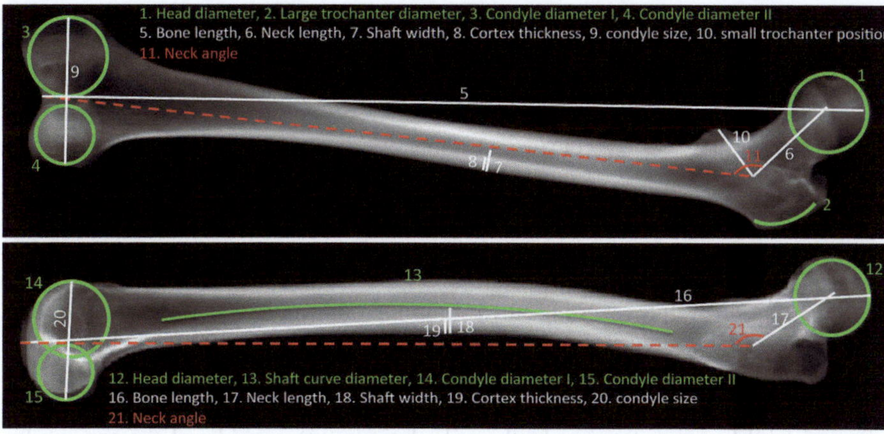

Fig. 10.2 The morphometric feature descriptors extracted from two orthogonal views. Diameters (in *green*), distances (in *white*), and angles (in *red*) are shown for frontal and lateral views. By selecting three landmarks for each *circle* fitting and two for each line we perform the measurements

both views. To model bone density information, the histogram of pixel intensities is calculated for the frontal view, generating a feature vector of size equal to the number of histogram bins.

From the triplets of (1) X-ray derived features, (2) patient's anthropometric data, and (3) corresponding bone shape and density scores, a random forest regression model is built. During testing, a new set of previously unseen X-ray orthogonal images is used to extract morphometric and bone density features to predict the bone shape and density scores. By cascading this model with the stress prediction model, described in the previous section, we are able to perform bone biomechanics *FEP* from X-ray images.

10.3 Results

In this section we show the results of the proposed method for fast FE predictions. First the database and tools used for the study are explained. It is followed by the results of FEP method for CT and X-ray scans. Database and Tools.

The database used in this study consists of 89 left femurs CT images. The resolution of CT scans was between 0.61 mm $\times$ 0.61 mm and 1.171 mm $\times$ 1.171 mm, with a slice thickness of 1 mm. The CT scans were acquired from 48 female and 41 male donors with average age, height, and weight of, respectively, 60.7 ± 16.2 years old, 165.70 ± 7.2 cm, and 70.1 ± 13.9 kg. Table 10.1 reports statistics about patients and femur morphometric in our database.

To study the accuracy of *FEP*, we used LOO [15] methodology to train with the maximum number of samples. The method was tested for one sample in the database

Table 10.1 The statistics of the bones used in the database ($n = 89$)

	Patients			Morphological parameters	
	Age	Height (cm)	Weight (kg)	Length (cm)	Anterior curve diameter (cm)
Min	23	150	42	37.8	57.0
Max	90	180	140	50.9	297.2
Mean	60.7	165.70	70.1	44.5	123.4
STD	16.2	7.2	13.9	23.2	33.4

when the rest of samples were used for training. This approach was repeated until every sample in the set was tested, which resulted in 89 different sets of training and testing samples.

For each training set, we built statistical models of shape and density [10], followed by FE computations. In the calculations, we considered the top modes of shape, density, and stress statistical models with the sum of more than 98 % of the variation in the dataset. As a result, 20 modes of shape and 46 modes of BMD were used for predicting the parameters of 16 modes of statistical model of stress. To tune the parameters of random forest, we performed a tenfold cross validation using the scikit-learn toolbox [16].

10.3.1 Results of FEP for CT Scans

We evaluated the prediction accuracy of stress values for each test sample. We calculated the correlation coefficient between the ground-truth stress values for each mesh node (as generated by the FE computations, using Abaqus FE solver in the normal walking loading situation) and the predicted Mises stress for those nodes. The average correlation coefficient for 89 test cases was 0.984 with a standard deviation of 0.008, showing the high accuracy of the proposed method.

We further evaluated the prediction performance by calculating the prediction error as the difference between ground-truth and predicted stress values. The ground-truth stress values, the predicted values, and the error distribution are shown in Fig. 10.3 for the best and the worst results. Among 89 samples, the best result was achieved with an average error (and standard deviation) of 0.058 (0.898) MPa in the mesh. For the worst case, the average error (and standard deviation) of the predicted stress values was equal to 1.316 (7.822) MPa. After examination of the worst-case result, we found that it corresponds to a patient with a body weight of 140 kg, while the maximum weight seen in training dataset was only 110 kg. This can be improved by using more samples for training to cover a larger variety of the population.

To evaluate *FEP* for different regions of interest, we also examined its accuracy in the femoral neck, femoral trochanter, and the femoral shaft, separately (see Fig. 10.4). The prediction error of stress for the neck region, which is the region of interest in fracture risk assessment, was smaller than 0.9 MPa in average.

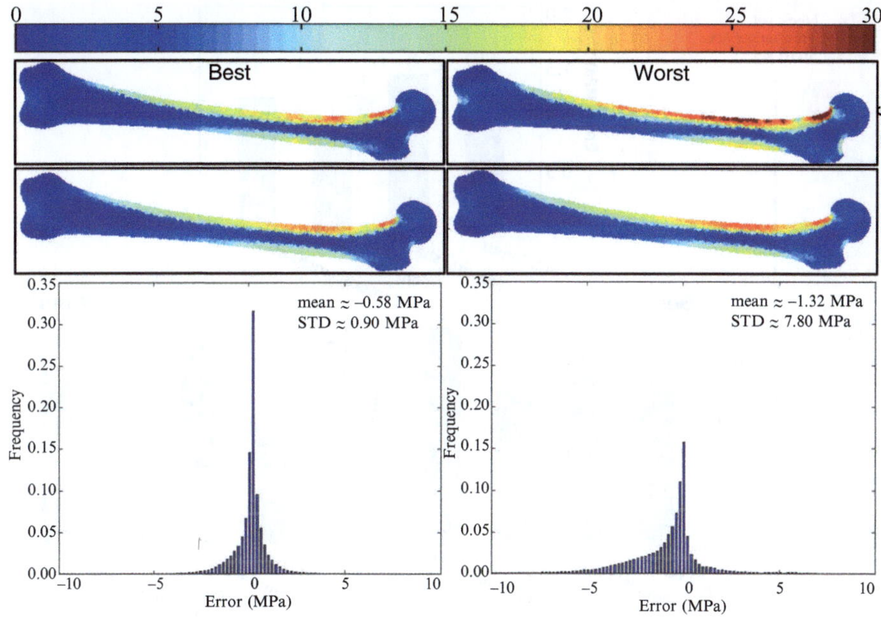

Fig. 10.3 The stress map predicted by FEP model for the best and the worst cases. From *top* to *bottom*: the stress map calculated using FE calculations (ground-truth), the predicted stress values for the corresponding bones, and the error distribution for these bones. The best result is achieved for the bone on the *left column* with a correlation coefficient of 0.994, and the worst prediction result in the database is in the *right column* with a correlation coefficient of 0.939. In absolute error distribution plots, we zoom in on the range of [−10, 10] MPa for better visibility. The frequency of error beyond this range is negligible (0.0003 and 0.0122, for the best and the worst case)

10.3.2 Results of FEP for X-ray Images

Based on features extracted from X-ray images we predicted the parameters of statistical shape and density models. We then used these parameters as input to our *FEP*. The average (standard deviation) correlation coefficient between the predicted stress using this method and the ground-truth values was 0.976 (0.012).

We developed a test case to evaluate the benefit of cascading two learning blocks (from X-ray to 3D data and from 3D data to stress parameters) as compared to a single learning model that directly predict stresses from X-ray based features. Similarly to the other models, the depth of trees is determined based on cross validation on training data. In this case the average correlation coefficient values dropped from 0.976 ± 0.012 to 0.956 ± 0.286. This shows that the cascading of two regression models, as proposed herein, does not significantly alter the accuracy of the prediction as compared to a single learning model. In addition, the cascading scheme has the extra value that other modality-specific models can be easily combined to *FEP*.

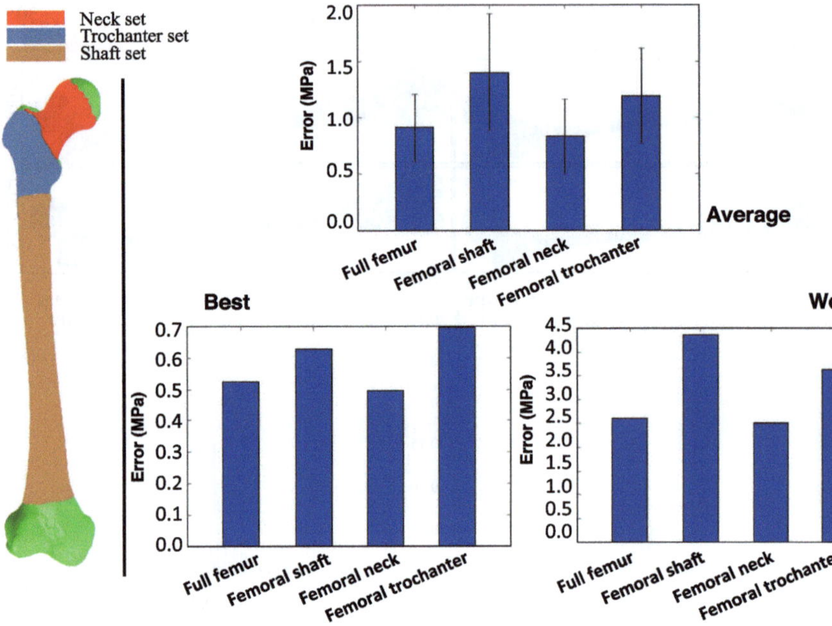

Fig. 10.4 The absolute error of *FEP* for different parts of the bone, (*top*) in average, (*bottom-left*) best case, and (*bottom-right*) for worst case. The different regions of interest are shown on the bone with different colors. *Red*: neck, *blue*: trochanter region, and *orange*: shaft

10.4 Conclusion and Discussion

It has been shown that using 3D FE analyses improves the osteoporosis diagnosis [4]. However clinical adoption of FE analysis in bone biomechanics and fracture risk assessment has been hampered by its computational complexity and required technical competences [5]. In this paper we developed a random-forest based regression framework to predict the results of the finite element prediction, termed here *FEP*, by simply selecting a couple of landmarks on clinical images. We proposed to use shape and density statistical model parameters to produce a compact and predictive set of features. In addition, the approach allows other image modalities to be used for prediction, and enables the incorporation of other emerging technologies developed for statistical shape modeling.

Using LOO experiments, comprising a database of 89 clinical cases, our method is capable of predicting the stress values for a walking loading condition with an average correlation coefficient of 0.984 and 0.976, for CT and X-ray images, respectively. These findings suggest that supervised learning approaches have the potential to leverage the clinical integration of mechanical simulations for the treatment of musculoskeletal conditions.

Motivated by the observed connections between the importance values obtained by random forest and actual models for shape, we analyzed the most important features in predicting the parameters of the stress statistical model. To predict the stress parameters, the body weight was found to be the most important parameter. This can be explained by the fact that in our experiments all bones were loaded in the walking situation when forces are scaled proportional to the body weight. Hence, the weight directly affects the stress values.

We note that our motivation is to demonstrate that even with rather simple, yet descriptive, selected features it is possible to yield a good level of prediction for bone biomechanics in real-time. In this work we used simple feature predictors for X-ray images. However, as proposed by the state-of-the-art approaches [17] three-dimensional bone shape and density parameters can be estimated robustly and accurately from X-ray images, which can further increase the predictive power of *FEP*.

Our method predicts the output stress values of an elastic material model for FE analysis from density and shape. However, it is flexible and can be easily adapted to incorporate more advanced mechanical parameters for predicting the bone fracture. For further improvement of *FEP*, we are planning to use existing methods on predicting trabecular bone structure from CT scans [18–20] to improve the estimation of biomechanical behavior of the bone.

This study has some limitations. First, the estimation of the scores of shape and density from CT scans was obtained using mesh registration. This registration task is time consuming and should be replaced by more effective methods such as active appearance model. However, this intermediate step is not necessary when the stress predictions are obtained from X-ray images. Another limitation results from the choice of synthetic images to mimic patients' X-ray images. This approach has been chosen to establish the method and avoid uncontrolled source of error. Clearly the accuracy of the predictions will decrease when clinical data will be used. Further studies will investigate this effect, but the high correlations reported in this study indicate that the prediction from clinical X-ray will provide accurate stress estimations. Finally, we observed that the method is not as successful in predicting the stress values for a bone of a patient who has the highest weight in our dataset. This problem occurs because no other patient with a similar body weight exists in our dataset. Similar to all other techniques that rely on machine learning, a large database that samples the population more evenly helps tackling this issue.

Our proposed approach followed by further improvements (adding trabecular bone structure to the analyses and using active shape modeling) shows a promising path towards real-time biomechanical analysis of bones in different patient-specific studies and brings an automated FE analysis to clinics. Since it is fast (the stress values are calculated in less than 1 s), several loading cases can be analyzed to have a better understanding of the patient's bone, moreover it can be used to find the bone strength and fracture risk for each individual patient.

The drawback of *FEP* is that for each loading case, i.e. walking, stance, side fall, one different model has to be trained. Note that this process is done offline during the training phase. The testing phase is fast and the stress values can be calculated in less than a second.

References

1. J.A. Kanis, *Assessment of Osteoporosis at the Primary Health Care Level* (World Health, 2007), p. 339
2. T. Whitmarsh, K.D. Fritscher, L. Humbert, L.M. Del Rio Barquero, T. Roth, C. Kammerlander et al., A statistical model of shape and bone mineral density distribution of the proximal femur for fracture risk assessment. Med. Image Comput. Comput. Assist. Interv. **14**, 393–400 (2011)
3. N. Sarkalkan, J.H. Waarsing, P.K. Bos, H. Weinans, A.A. Zadpoor, Statistical shape and appearance models for fast and automated estimation of proximal femur fracture load using 2D finite element models. J. Biomech. **47**, 3107–3114 (2014). doi:10.1016/j.jbiomech.2014.06.027
4. E. Dall'Ara, B. Luisier, R. Schmidt, F. Kainberger, P. Zysset, D. Pahr, A nonlinear QCT-based finite element model validation study for the human femur tested in two configurations in vitro. Bone **52**, 27–38 (2013). doi:10.1016/j.bone.2012.09.006
5. S. Poelert, E. Valstar, H. Weinans, A.A. Zadpoor, Patient-specific finite element modeling of bones. Proc. Inst. Mech. Eng. H **227**, 464–478 (2013). doi:10.1177/0954411912467884
6. T. Heimann, H.-P. Meinzer, Statistical shape models for 3D medical image segmentation: a review. Med. Image Anal. **13**, 543–563 (2009). doi:10.1016/j.media.2009.05.004
7. L. Grassi, E. Schileo, C. Boichon, M. Viceconti, F. Taddei, Comprehensive evaluation of PCA-based finite element modelling of the human femur. Med. Eng. Phys. **36**, 1246–1252 (2014). doi:10.1016/j.medengphy.2014.06.021
8. I. Castro-Mateos, J.M. Pozo, T.F. Cootes, J.M. Wilkinson, R. Eastell, A.F. Frangi, Statistical shape and appearance models in osteoporosis. Curr. Osteoporos Rep. **12**, 163–173 (2014). doi:10.1007/s11914-014-0206-3
9. T.F. Cootes, C.J. Taylor, D.H. Cooper, J. Graham, Active shape models-their training and application. Comput. Vis. Image Underst. **61**, 38–59 (1995). doi:10.1006/cviu.1995.1004
10. S. Bonaretti, C. Seiler, C. Boichon, M. Reyes, P. Büchler, Image-based vs. mesh-based statistical appearance models of the human femur: implications for finite element simulations. Med. Eng. Phys. **36**, 1626–1635 (2014). doi:10.1016/j.medengphy.2014.09.006
11. C. Boichon, et al., Shape indexing of human femur using morphing and principal component analysis. VPH (2010)
12. A.D. Speirs, M.O. Heller, G.N. Duda, W.R. Taylor, Physiologically based boundary conditions in finite element modelling. J. Biomech. **40**, 2318–2323 (2007). doi:10.1016/j.jbiomech.2006.10.038
13. L. Breiman, Random forests. Mach. Learn. **45**, 5–32 (2001). doi:10.1023/A:1010933404324
14. A. Criminisi, J. Shotton (eds.), *Decision Forests for Computer Vision and Medical Image Analysis*, vol. 1 (Springer, London, 2013). doi:10.1007/978-1-4471-4929-3
15. S. Geisser, *Predictive Inference* (CRC Press, 1993)
16. F. Pedregosa, G. Varoquaux, A. Gramfort, V. Michel, B. Thirion, O. Grisel et al., Scikit-learn: machine learning in Python. J. Mach. Learn. Res. **12**, 2825–2830 (2011)
17. G. Zheng, 3D volumetric intensity reconstruction from 2D X-ray images using partial least squares regression, in *2013 IEEE 10th International Symposium on Biomedical Imaging* (IEEE, 2013), pp. 1268–1271. doi:10.1109/ISBI.2013.6556762
18. E. Taghizadeh, G. Maquer, M. Reyes, P. Büchler, *Including the Trabecular Anisotropy from Registered microCT Data in Homogenized FE Model Improves the Bone's Mechanical Predictions* (CMBBE, Amsterdam, 2014)
19. D. Larsson, B. Luisier, M.E. Kersh, E. Dall'ara, P.K. Zysset, M.G. Pandy et al., Assessment of transverse isotropy in clinical-level CT images of trabecular bone using the gradient structure tensor. Ann. Biomed. Eng. **42**, 950–959 (2014). doi:10.1007/s10439-014-0983-y
20. K. Lekadir, J. Hazrati-Marangalou, C. Hoogendoorn, Z. Taylor, B. van Rietbergen, A.F. Frangi, Statistical estimation of femur micro-architecture using optimal shape and density predictors. J. Biomech. **48**, 598–603 (2015). doi:10.1016/j.jbiomech.2015.01.002

Chapter 11
Some Use Cases for Composite Finite Elements in Image Based Computing

Lars Ole Schwen, Torben Pätz, and Tobias Preusser

11.1 Introduction

In the past decades mathematical modeling, simulation, and optimization have become indispensable tools in systems biology, systems medicine, as well as medical diagnosis and treatment-planning. In the "image based computing" paradigm, radiological images like CT, MRI, ultrasound, etc. are analyzed to yield segmented structures of organs, tissue, or other structures pictured. The consequent goal is to simulate physiological processes, or to simulate and optimize treatments using mathematical models and their numerical implementations. A particular challenge in this is, however, the generation of computational meshes from the segmented imaging data that is needed in the process of discretization of models.

In fact, structures in organisms have a complicated geometry. They are in general irregularly shaped and show large intra- and inter-individual variations. Moreover, it is often necessary to also resolve internal sub-structures or interfaces with a computational mesh, thus to account for various bio-physical properties of the many tissue-types that might be involved and which may be discontinuous at the internal tissue interfaces. These facts make mesh generation for image based computing a difficult task that, moreover, must respect constraints of a clinical workflow in case of a true medical application in daily routine.

For the discretization of mathematical models that are characterized by partial differential equations (PDEs) computational meshes are directly related to the Finite

L.O. Schwen (✉) • T. Pätz
Fraunhofer MEVIS, Bremen, Germany
e-mail: ole.schwen@mevis.fraunhofer.de; torben.paetz@mevis.fraunhofer.de

T. Preusser
Fraunhofer MEVIS, Bremen, Germany

Jacobs University Bremen, Bremen, Germany
e-mail: tobias.preusser@mevis.fraunhofer.de

G. Joldes et al. (eds.), *Computational Biomechanics for Medicine*,
DOI 10.1007/978-3-319-28329-6_11

Element spaces spanning the space of solutions. In the case of irregular boundaries or internal structures, it is known that the solutions have less regularity that may result, e.g., in kinks and discontinuous gradients. The straightforward approach to tackle these irregularities is mesh adaptivity, i.e., decreasing the size of the mesh's cells in areas with lower regularity of the solution and thus adapting the associated Finite Element spaces. Other approaches avoiding mesh adaptivity that have been discussed in the literature include generalized FEM (GFEM), extended FEM (XFEM), immersed FEM, fictitious domain methods, WEB-splines, and others. We refer to [18, 20, 26] for literature overviews and [3, 6, 7, 12, 16, 17, 25] for selected more recent approaches.

In this paper we review the Composite Finite Element (CFE) approach to image based computing. The CFE approach goes back to [9, 24]. This paper is a summary of our work from the past decade originally published in [18, 20, 21, 23, 26–30, 35]. The method works efficiently on regular hexahedral grids as they are provided by the usual three-dimensional voxel grids of medical images. Still it allows for the resolution of complicated geometries and interfaces by automatically adapting standard linear Finite Element basis functions and thus modifying the corresponding Finite Element space accordingly. In the remainder of the paper we first explain the general idea of CFE in more detail in Sect. 11.2 and an approach for numerical homogenization in Sect. 11.3. Then, in Sect. 11.4, we show two applications from the field of medical image computing that demonstrate the use of CFE: the simulation of radio-frequency ablation (RFA), the simulation of the elastic deformation of vertebral trabecular bone, and numerical homogenization for the latter. We close the paper with a summary and conclusions in Sect. 11.5.

11.2 Composite Finite Elements

We will describe the concept of CFE for the domain $\Omega = (0, 1)^3$. The domain will be discretized with a regular hexahedral grid $\mathcal{G}$ that has 2^{3l} elements of grid width $h = 2^{-l}$ and a total of $(2^l + 1)^3$ nodes. We choose to work with the unit cube here as it eases the presentation. The application of CFE to other cuboid domains is of course possible straightforwardly. Working with piecewise affine-linear basis functions, we subdivide each hexahedron in six tetrahedra in such a way that edges are consistent with neighboring elements, resulting in the mesh $\mathcal{G}^{\boxtimes}$ denoted as the regular tetrahedral mesh.

We assume that the domain Ω contains $\Omega_i \subset \Omega$, the object we are interested in, and which has a complicated boundary. Consequently the solution to an elliptic (or parabolic) PDE will be supported on Ω_i and we will build a Finite Element space whose basis functions are supported on Ω_i. If in addition the object comprises complex interfaces between materials of different bio-physical properties, the solution to the PDE will have kinks at the interface. Such kinks result from discontinuities of material properties when they have different values on both sides of the interface. The material properties enter the equations as coefficients (e.g., as diffusivities or elasticity parameters) and the kink in the solution will depend on the ratio of the material property across the interface. Again, in CFE we will build basis functions that are able to interpolate functions fulfilling the kink condition. This approach to complex object boundaries and interfaces makes classical grid adaptivity obsolete.

To proceed let us assume that $\overline{\Omega} = \overline{\Omega_i \cup \Omega_e}$ and further $\overline{\Omega_i} = \overline{\Omega_+ \cup \Omega_-}$ where $\Omega_\pm$ are disjoint sets. Thus, the domain Ω is decomposed into the object Ω_i and its exterior Ω_e. The object Ω_i contains two material domains $\Omega_\pm$. A generalization to more objects and more materials is of course easily possible [27]. In image based computing it is convenient to define these domains from 3D image data that is provided on a regular hexahedral voxel grid. With image processing methodology level-set functions can be provided such that the zero levelsets define the interfaces of the domains, see below.

In the following we will describe how to construct CFE for complicated domains, for complicated interfaces between different materials, and how to use CFE in the context of homogenization. Note that our expositions will be brief and just explaining the principal concepts. For more details we refer the reader to the original publications that mentioned the respective sections below.

The treatment of different cases of boundary conditions (Dirichlet and Neumann; on the bounding box and on the interface; zero and nonzero) is addressed in [27]. A key advantage of the underlying uniform hexahedral grids is their natural hierarchy of coarse scales. These were used in a CFE multigrid solver for the case of complicated domains [18]. Defining a suitable coarsening scheme for CFE for discontinuous coefficients turned out to be challenging [20] and requires further investigation. One possibility could be a hybrid approach combining standard geometric, algebraic [33], and topological [5] coarsening.

11.2.1 CFE for Complicated Domains

Let us first consider the case of a domain with complicated boundary consisting of only a single material, w.l.o.g. described by $\Omega_+ = \emptyset$ and $\Omega_i = \Omega_-$. The interior boundary is then given as $\Gamma = \partial\Omega_i \cap \Omega$. Let $\varphi\colon \Omega \to \mathbb{R}$ be the level-set function defining this domain, i.e., a function whose zero sub-levelset is Ω_-.

In this case, CFE basis functions are constructed to be standard affine FE basis functions restricted to Ω_i as shown in Fig. 11.1 for the 2D case. This construction is achieved by introducing an auxiliary mesh $\mathcal{G}^\triangle$. For this purpose, tetrahedra of $\mathcal{G}^\boxtimes$ intersected by the interior boundary Γ are further subdivided into four or six sub-tetrahedra such that the a linear approximation to the boundary Γ is resolved. In fact, the auxiliary nodes $n_i^\triangle$ needed for the construction of the auxiliary mesh are computed as the zero crossings of the affine-linear interpolation of φ on the edges of $\mathcal{G}^\boxtimes$ that are intersected by Γ. From the standard, piecewise affine basis functions $\mathcal{G}^\triangle$, CFE basis functions are composed as a weighted sum, where the weights are given by the barycentric coordinates of the auxiliary nodes $n_i^\triangle$ on the respective edges of $\mathcal{G}^\boxtimes$.

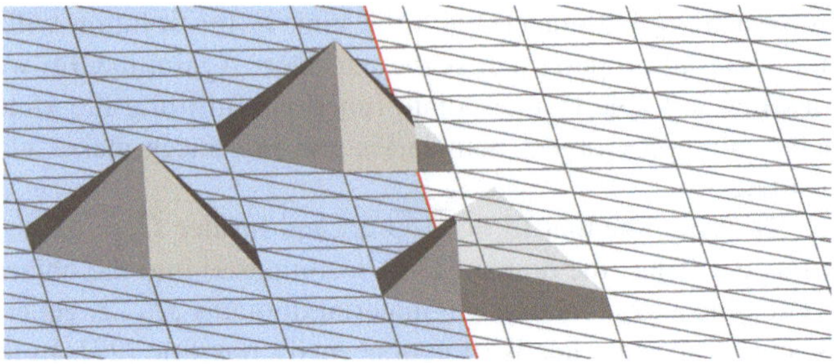

Fig. 11.1 2D CFE basis functions for complicated domains. For the domain (*left*, *light blue region*), CFE basis functions in the interior are standard piecewise affine tent functions for the nodes of the regular tetrahedral mesh $\mathcal{G}^{\boxtimes}$. For the exterior (*right*, *white region*), there are no degrees of freedom and no basis functions. Near the interface (*red line*), standard tent functions are restricted to the interior of the domain. This applies to the nodes of $\mathcal{G}^{\boxtimes}$ adjacent to the intersection of the interface (figure from [26, Fig. 3.11])

For a detailed discussion of the CFE construction including the description of a multigrid solver, we refer to [18, 26].

11.2.2 CFE for Discontinuous Coefficients

For the case of discontinuous coefficients, let us assume that $\Omega = \Omega_i$, i.e., there is no additional complicated domain boundary. In this case, let $\Gamma = \partial\overline{\Omega_+} \cap \partial\overline{\Omega_-}$, i.e., the interface between the two different materials, and let again $\varphi\colon \Omega \to \mathbb{R}$ be the function defining this interface, i.e., φ is positive/negative in $\Omega_\pm$ and Γ is its zero levelset.

In this case, CFE basis functions are constructed such that they are capable of interpolating functions satisfying the kink condition due to the parameter discontinuity (denoted below as "admissible" functions). The construction starts similar to the case above, introducing $\mathcal{G}^{\boxtimes}$ which now approximates the interface between two material domains. Next, composition weights need to be determined to obtain CFE basis functions as weighted sums of standard, piecewise affine basis functions $\mathcal{G}^{\triangle}$. For this purpose, we consider the problem of locally interpolating admissible functions at nodes $n_i^{\triangle}$ of the auxiliary mesh $\mathcal{G}^{\triangle}$ from nodes $n_j^{\boxtimes}$ of $\mathcal{G}^{\boxtimes}$. The admissibility condition (= kink condition) involves the local interface geometry, the underlying PDE, and the values of its coefficients on both sides of the interface. Using a Taylor expansion, known properties can be exploited, namely continuity of (a) the function, of (b) its derivative in tangential directions, and of (c) coefficient times derivative in normal direction of the function. Averaging over all adjacent tetrahedra of $\mathcal{G}^{\boxtimes}$ provides the interpolation weight $w_{i,j}$ for the pair $\left(n_i^{\triangle}, n_j^{\boxtimes}\right)$. In turn, $w_{i,j}$ is finally used as the composition weight with which the auxiliary basis function for $n_i^{\triangle}$ of $\mathcal{G}^{\triangle}$ contributes to the CFE basis function for $n_j^{\boxtimes}$ of $\mathcal{G}^{\boxtimes}$.

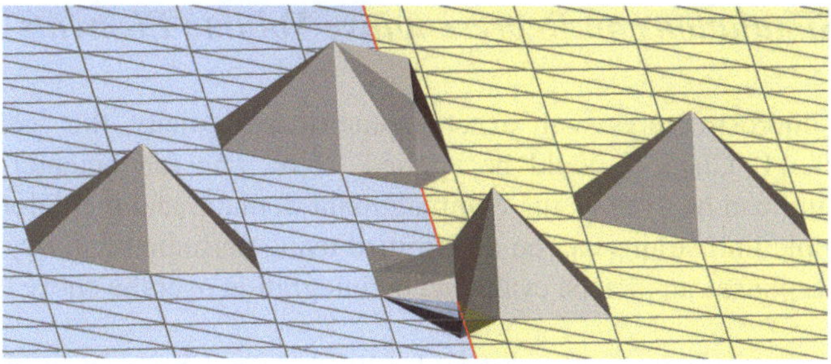

Fig. 11.2 2D CFE basis functions for complicated interfaces. For piecewise constant coefficients in two domains (*light blue* and *yellow regions*) with a discontinuity across the interface (*red line*), CFE basis functions far from the interface are again standard piecewise affine tent functions for the nodes of the regular tetrahedral mesh $\mathcal{G}^{\boxtimes}$. At the interface, CFE basis functions are constructed in such a way that they can interpolate the local kink condition, in this case for isotropic scalar coefficients with a ratio of 1 : 10. These CFE basis functions at the interface may attain values outside [0, 1] and may have extended support. Still, they form a partition of unity and their support remains local and bounded (figure from [26, Fig. 3.12])

For diffusion as a scalar model problem, a 2D example is shown in Fig. 11.2. The resulting basis functions remain a partition of unity and retain boundedness of their supports. However, they may attain values below 0 (see Fig. 11.2) or greater than 1 near the interface.

In case of vector-valued elasticity with discontinuous material parameters, the construction is more technical as the coupling condition at the interface then involves three dimensions simultaneously. For nodes near the interface, this leads to three CFE basis functions, each of which has contributions in all three space dimensions. For a more detailed description of the construction for the scalar and vector-valued model problems, we refer to [20, 26].

11.3 Numerical Homogenization with CFE

Being specifically designed for simulations on cuboid domains, CFE are well suited for numerical homogenization, i.e., for determining effective macroscopic material parameters. An approach for numerical homogenization for linear elasticity of trabecular bone specimens was presented in [23], other approaches include [8, 13]. The basic idea of our approach [23] is to simulate six cases of uniaxial compression and shearing ("macroscopic unit strains") and determine the respective stress response of a statistically representative cubic part of the trabecular bone. Together, this provides the necessary information for the macroscopic linear elasticity tensor.

11.3.1 Simulations on Representative Volume Elements

In case of microstructures with exact geometric periodicity, the "cell problem" approach cf. [1, Chap. 1] could be applied. Here, one fundamental cell of the microstructure can be used as the computational domain, and the deformation due to macroscopic unit strains can be determined using periodic boundary conditions. The stress response can then be evaluated by integration over the entire fundamental cell.

As biological structures like trabecular bone do not have an exactly periodic geometric structure, there is no geometrically fundamental cell of the object on which to impose periodic boundary conditions. Thus, we modify the "cell problem" approach and use a statistically representative part of the structure as the computational domain, which we call representative volume element or RVE [14]. For simplicity, the RVE is assumed to be cube-shaped. Macroscopic unit strains are imposed by corresponding Dirichlet boundary conditions on the entire outer boundary of the RVE. Evaluating the stress response is then restricted to an inner part of the RVE sufficiently far from the outer boundary. This is necessary to avoid artificial stiffening due to boundary effects [20, Fig. 7.3]. A suitable thickness for the boundary layer omitted for the stress evaluation turned out to be [20, Sect. 7.2] approximately one eighth of the edge length of the RVE. Based on [10, 34], the size of the interior used for stress evaluation was chosen to be at least five times the intertrabecular distance (pore size).

11.3.2 Macroscopic Linear Elasticity Tensors

From the approach described above, the stress response for each unit strain is obtained, providing all parameters of the macroscopic elasticity tensor, i.e., the effective elasticity tensor.

For these, the question arises whether they are approximately orthotropic, and if so, what is the orientation of the axes. For this purpose, we solve an optimization problem, finding the rotation of the coordinate system for which the deviation of the tensor from an orthotropic one is minimized. The objective function representing this deviation is obtained from the tensor written in Voigt's notation where orthotropy is observed by certain entries being zero. Again, we refer the reader to [23] for more details on the approach.

11.4 Applications

In the following we will briefly describe some use cases for the CFE approaches described so far.

11.4.1 Vaporization During Radio-Frequency Ablation

As a first application reported in [19], we consider a simulation of RFA. RFA is a minimally invasive technique for the treatment of lesions, e.g., liver cancer [2, 15]. For RFA, a thin probe carrying electrodes is placed percutaneously inside the tumor and connected to an electric generator. Due to the electric resistance of the tissue, heat develops and destroys proteins and thus cells. When the temperature exceeds 100 °C, the water inside the tissue vaporizes. This changes both the heat conductivity and the electric conductivity, and thus the electric potential causing the heating.

To simulate the temperature evolution during RFA, we couple models for electrostatic fields and temperature diffusion with a model for phase changes to account for the water evaporization. Thus, we have a free boundary problem with a complicated boundary that is coupled with discontinuous material coefficients across this interface. In the following, we discretize this problem with the CFE approach. In the notation introduced above, we consider the liquid phase to be "the object" $\Omega_i = \Omega_l$ and the gaseous phase to be "the exterior" $\Omega_e = \Omega_g$. Here and in the following, indices l denote quantities in the liquid phase, whereas the subscript g denotes quantities in the gaseous phase.

Phase Change The phase change is modeled by a well-known Stefan problem [31] that describes the discontinuity of the temperature gradient across an interface between water in different phases, e.g., between liquid and vaporized water:

$$\frac{[-\lambda \nabla T \cdot N]}{L} = \rho_l \Big(v_l - D \Big), \tag{11.1}$$

where λ is the thermal conductivity, T the temperature, N the unit normal to the interface pointing from the liquid to the vapor domain, L the latent heat of the phase change, ρ_l the density of the liquid, v_l the liquid vapor speed at the interface, and D the interface speed in normal direction. The jump operator $[A] := A_g - A_l$ denotes the difference between quantities on the liquid and vapor side of the interface. The Stefan condition allows to decouple the heat transfer equations in the liquid and vapor phase. Thus, for the heat diffusion we end up with the equations

$$\begin{aligned} \rho_g c_g \partial_t T + \rho_g c_g V \cdot \nabla T &= -\mathrm{div}(\lambda_g \nabla T) + Q_{\mathrm{rf}} \quad \text{in } \Omega_g \times \mathbb{R}^+, \\ \rho_l c_l \partial_t T &= -\mathrm{div}(\lambda_l \nabla T) + Q_{\mathrm{rf}} \quad \text{in } \Omega_l \times \mathbb{R}^+, \end{aligned} \tag{11.2}$$

with appropriate initial and boundary conditions and where Q_{rf} is the heat source according to the electric field caused by the RF current, see below. These equations with complicated shaped domain boundary on the liquid–vapor boundary are discretized and solved using the CFE approach presented above. For details, we refer to [19].

RFA Simulation The second main component of the RFA simulation is the solution of the electrostatic equation

$$-\text{div}(\sigma \nabla \Phi) = 0 \text{ in } \Omega \tag{11.3}$$

with appropriate boundary conditions. This equation provides the electric potential and thus the heat source $Q_{\text{rf}} = \sigma \|\nabla \Phi\|^2$. Here, σ is the electric conductivity that has a discontinuity at the interface separating liquid and gaseous domain.

Thus, for the overall RFA simulation, we need to solve (a) two heat transfer equations on the vapor and liquid domains of complicated shape, one for each phase, and (b) the potential equation for computing the electric potential with a discontinuous coefficient on the interface between the two domains. The evolution of the interface is obtained through the Stefan condition from above.

With this RFA model, we were able to calculate the expansion of a vapor bubble around the probe (see Fig. 11.3). With our simulation we achieve results comparable to measurements from ex situ experiments. Additionally, we computed the impedance during ablation in our simulations and compared them to measurements from [32, Fig. 7-1], see Fig. 11.4. Indeed, the characteristics of the curves for numerical simulation and experiment coincide, i.e., the impedance slightly decreases at the beginning; it rises steeply when the vapor bubble around the applicator is established; and the impedance remains constant afterwards.

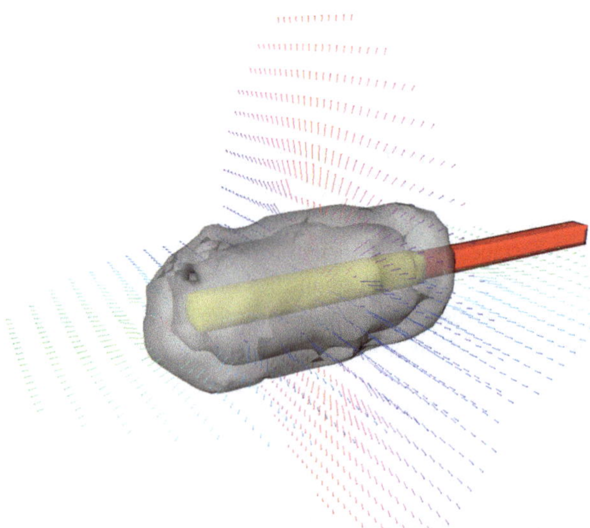

Fig. 11.3 Expansion of the vapor phase around an RF probe. *Arrows*, *color* coded by velocity, indicate the vector field that drives the evolution of the interface, i.e., the water bubble (image from [19, Fig. 4.6]. Copyright ©2012 Society for Industrial Mathematics. Reprinted with permission. All rights reserved.)

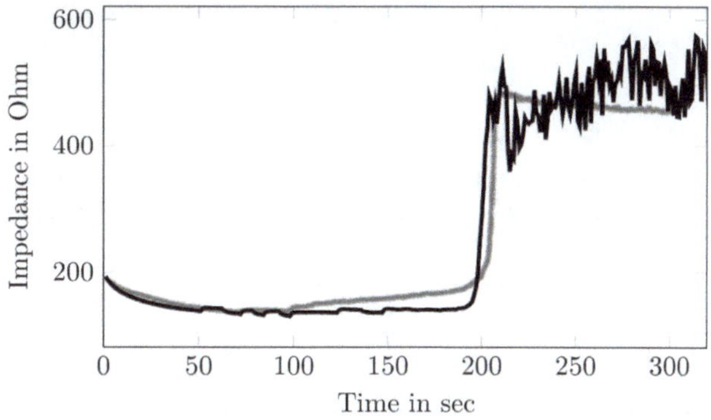

Fig. 11.4 Comparison of the impedance measured during an ablation (*grey curve*, experimental data from [32]) and computed from the simulation (*black curve*) (figure adapted from [19, Fig. 4.7]. Original image Copyright ©2012 Society for Industrial Mathematics. Reprinted with permission. All rights reserved.)

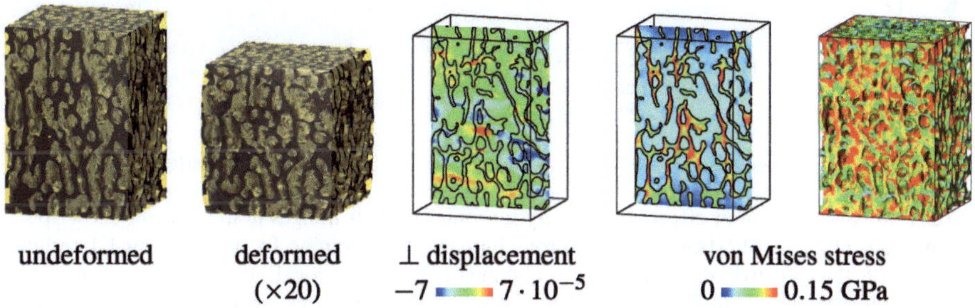

Fig. 11.5 Vertebroplasty. For a porcine trabecular bone specimen virtually embedded in Polymethylmethacrylate (PMMA), compression was simulated. On a slice through the center of the specimen, the displacements perpendicular to the slice are shown in units relative to the specimen height. Moreover, the von Mises stress on that slice as well as the bone/PMMA interface is visualized. These visualizations show the impact of the parameter discontinuity across the interface of geometrically complicated shape (figure adapted from [26, Fig. 7.28] and [20, Fig. 6.5])

11.4.2 Elastic Deformation of Trabecular Bone

Vertebroplasty As an application involving linear elasticity with discontinuous coefficients, we consider an example from [20, 26], a specimen of a porcine T1 vertebral body virtually embedded in Polymethylmethacrylate (PMMA) subject to 1 % longitudinal compression. Material properties for the bone are assumed to be those for human vertebral bodies, $E = 13\,\text{GPa}$ and $\nu = 0.32$ [36], and $E = 3\,\text{GPa}$ and $\nu = 0.38$ for PMMA as in [20]. With an isotropic image resolution of $35\,(\mu)\text{m}$, the resulting computational mesh had $143 \times 143 \times 214$ nodes. Results of this simulation are shown in Fig. 11.5.

Effective Elasticity Tensors of Trabecular Bone in Different Species As a second application involving trabecular bone, we investigate differences in the macroscopic stiffness of specimens taken from vertebrae of different species [23]. For this purpose, cubic specimens of edge length 5.16 mm were obtained from a young male human, an osteoporotic female human, a porcine, and a bovine spine. The values above, $E = 13\,\text{GPa}$ and $\nu = 0.32$, were used as material properties for

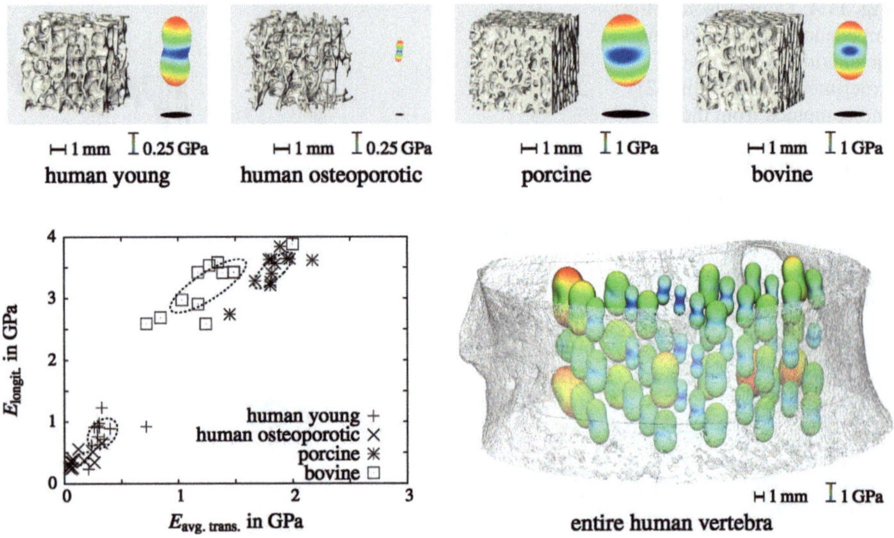

Fig. 11.6 Comparison of effective elasticity tensors. The four images in the *top row* show specimens from a human T12, from an osteoporotic human T10, a porcine T1, and a bovine L1 vertebra along with the effective elasticity tensors obtained by our numerical homogenization procedure (images adapted from [26, Fig. 7.36]). The overall stiffness matches the visual impression of the bone density, and the stiffness is largest in approximately the craniocaudal direction, corresponding to the vertical axis in the specimens. The *bottom left* plot provides a comparison of longitudinal stiffness and average stiffness in the transverse directions for multiple specimens of the species above (adapted from [23, Fig. 2]). This shows a rather clear clustering in terms of both absolute stiffness and ratio of longitudinal over transverse stiffness. The *bottom right* image comprises the effective elasticity tensors obtained for different positions within an entire human L4 vertebra (image adapted from [29, Fig. 6], original image Copyright ©2012 Taylor & Francis, http://www.tandfonline.com/), possibly with special formatting for the URL?

the trabecular bone, this time viewed as a complicated domain without surrounding medium. Using the numerical homogenization approach above, effective stiffness tensors were obtained for each of the specimens, visualized in Fig. 11.6.

Here, in order to give a quick visual impression of the macroscopic elasticity properties, we use the visualization presented in [4, 11]: A sphere is deformed according to the compressive stiffness in different directions, and it is rendered colored according to the respective bulk modulus, resulting in the colored "peanuts" shown in Fig. 11.6.

Plotting the longitudinal and the average transverse stiffness for all specimens, a clear clustering of the different species can be observed. In [29], intravertebral variations were investigated further for multiple samples of an entire human vertebra, also shown in Fig. 11.6.

A validation of the CFE elasticity simulation and the homogenization has been discussed in [29]. There, local stiffness tensors have been computed for the trabecular core of a female human lumbar vertebra (58 years) which had been scanned by μCT. Figure 11.7 briefly summarizes the results in terms of a correlation

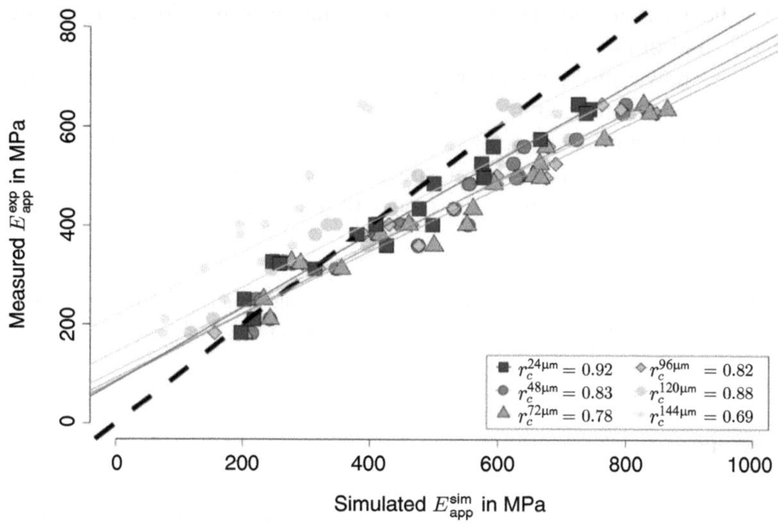

Fig. 11.7 Correlations between experiment and CFE homogenization. Correlation lines for resolutions below 168 (μ)m were statistically significant ($p < 0.05$). Concordance coefficients r_c are given in the *box* for the varying resolutions. Ideal concordance would have been given by a correlation straight falling together with the *black, dashed line* (image adapted from [29, Fig. 3], original image Copyright ©2012 Taylor & Francis, http://www.tandfonline.com/, again possibly with special formatting for the URL?)

plot for measured and experimental apparent stiffness for various resolutions of the image data that shows the trabecular bones. Our investigations show that there is a moderate but acceptable agreement between experiment and numerical homogenization.

11.5 Conclusions

We have discussed a CFE approach to image based computing. The CFE method is capable of resolving complicated structures or interfaces on hexahedral grids as they are provided by standard voxel grids of three-dimensional medical image data. In the paper we have presented possible use cases for CFE, when the domain has a complicated boundary, when an interface between different materials has complicated shape, or when the computation of efficient macroscopic quantities, i.e., homogenization of complicated materials, is of interest. In contrast to standard FE on hexahedral grids, approximate interfaces are resolved. Thus, a higher order of convergence by CFE is expected and was verified in [19, 20, 27].

The method is based on a virtual sub-division of the hexahedral grids into tetrahedra. The CFE discretization can be implemented efficiently and a multigrid solver has been developed. The CFE implementation from [26] is open source and available as part of the QuocMesh software library [22].

We have shown the application of the CFE discretization to the simulation of radio-frequency ablation in which the electric conductivity is discontinuous across a moving interface that has complicated shape. Also we demonstrate the homogenization approach for the elastic deformation of trabecular bone. As said before, for more details we refer the reader to the original publications mentioned above.

Acknowledgements We acknowledge Martin Rumpf, Stefan Sauter, and Uwe Wolfram for their collaboration and many fruitful and inspiring discussions regarding CFE and their applications.

References

1. G. Allaire, *Shape Optimization by the Homogenization Method*. Applied Mathematical Sciences, vol. 146 (Springer, New York, 2002)
2. E.J. Berjano, Theoretical modeling for radiofrequency ablation: state-of-the-art and challenges for the future. BioMed. Eng. OnLine **5**, 24 (2006)
3. A. Bonito, R.A. DeVore, R.H. Nochetto, Adaptive finite element methods for elliptic problems with discontinuous coefficients. SIAM J. Numer. Anal. **51**(6), 3106–3134 (2013)
4. A. Cazzani, M. Rovati, Extrema of Young's modulus for cubic and transversely isotropic solids. Int. J. Solids Struct. **40**, 1713–1744 (2003)
5. C. Dick, J. Georgii, R. Westermann, A real-time multigrid finite hexahedra method for elasticity simulation using CUDA. Simul. Model. Pract. Theory **19**, 801–816 (2011)
6. Y. Efendiev, J. Galvis, T.Y. Hou, Generalized multiscale finite element methods (GMsFEM). J. Comput. Phys. **251**, 116–135 (2013)
7. S. Frei, T. Richter, A locally modified parametric finite element method for interface problems. SIAM J. Numer. Anal. **52**(5), 2315–2334 (2014)
8. L.J. Gibson, Biomechanics of cellular solids. J. Biomech. **38**(3), 377–399 (2005)
9. W. Hackbusch, S.A. Sauter, Composite finite elements for the approximation of PDEs on domains with complicated micro-structures. Numer. Math. **75**(4), 447–472 (1997)
10. T.P. Harrigan, M. Jasty, R.W. Mann, W.H. Harris, Limitations of the continuum assumption in cancellous bone. J. Biomech. **21**(4), 269–275 (1988)
11. Q.C. He, A. Curnier, A more fundamental approach to damaged elastic stress-strain relations. Int. J. Solids Struct. **32**(10), 1433–1457 (1995)
12. J.L. Hellrung Jr., L. Wang, E. Sifakis, J.M. Teran, A second order virtual node method for elliptic problems with interfaces and irregular domains in three dimensions. J. Comput. Phys. **231**(4), 2015–2048 (2012)
13. S.J. Hollister, D.P. Fyhrie, K.J. Jepsen, S.A. Goldstein, Application of homogenization theory to the study of trabecular bone mechanics. J. Biomech. **24**(9), 825–839 (1991)
14. S.J. Hollister, N. Kikuchi, A comparison of homogenization and standard mechanics analyses for periodic porous composites. Comput. Mech. **10**(2), 73–95 (1992)
15. T. Kröger, I. Altrogge, T. Preusser, P.L. Pereira, D. Schmidt, A. Weihusen, H.O. Peitgen, Numerical simulation of radio frequency ablation with state dependent material parameters in three space dimensions, in *MICCAI (2)*, ed. by R. Larsen, M. Nielsen, J. Sporring. Lecture Notes in Computer Science, vol. 4191 (Springer, New York, 2006), pp. 380–388
16. G. Legrain, P. Cartraud, I. Perreard, N. Moës, An X-FEM and level set computational approach for image-based modelling: application to homogenization. Int. J. Numer. Methods Eng. **86**(7), 915–934 (2011)

17. X. Li, J. Lowengrub, A. Rätz, A. Voigt, Solving PDEs in complex geometries: a diffuse domain approach. Commun. Math. Sci. **7**(1), 81 (2009)
18. F. Liehr, T. Preusser, M. Rumpf, S. Sauter, L.O. Schwen, Composite finite elements for 3D image based computing. Comput. Vis. Sci. **12**(4), 171–188 (2009)
19. T. Pätz, T. Preusser, Composite finite elements for a phase change model. SIAM J. Sci. Comput. **34**(5), B672–B691 (2012)
20. T. Preusser, M. Rumpf, S. Sauter, L.O. Schwen, 3D composite finite elements for elliptic boundary value problems with discontinuous coefficients. SIAM J. Sci. Comput. **33**(5), 2115–2143 (2011)
21. T. Preusser, M. Rumpf, L.O. Schwen, Finite element simulation of bone microstructures, in *Proceedings of the 14th Workshop on the Finite Element Method in Biomedical Engineering, Biomechanics and Related Fields*, University of Ulm, July 2007, pp. 52–66
22. A.G. Rumpf, Institute for Numerical Simulation, University of Bonn: Quocmesh software library, (2011). http://numod.ins.uni-bonn.de/software/quocmesh/index.html
23. M. Rumpf, L.O. Schwen, H.J. Wilke, U. Wolfram, Numerical homogenization of trabecular bone specimens using composite finite elements, in *The International Journal of Multiphysics, Special Edition: Multiphysics Simulations – Advanced Methods for Industrial Engineering. Selected Contributions from 1st Fraunhofer Multiphysics Conference*, 2010, pp. 127–143
24. S.A. Sauter, R. Warnke, Composite finite elements for elliptic boundary value problems with discontinuous coefficients. Computing **77**(1), 29–55 (2006)
25. D. Schillinger, E. Rank, An unfitted *hp*-adaptive finite element method based on hierarchical B-splines for interface problems of complex geometry. Comput. Methods Appl. Mech. Eng. **200**(47), 3358–3380 (2011)
26. L.O. Schwen, Composite finite elements for trabecular bone microstructures. Ph.D. thesis, University of Bonn, 2010
27. L.O. Schwen, T. Pätz, T. Preusser, Composite finite element simulation of radio frequency ablation and bone elasticity, in *Proceedings of the 6th European Congress on Computational Methods in Applied Sciences and Engineering (ECCOMAS 2012)*, Vienna, ed. by J. Eberhardsteiner, et al., September 2012
28. L.O. Schwen, T. Preusser, M. Rumpf, Composite finite elements for 3D elasticity with discontinuous coefficients, in *Proceedings of the 16th Workshop on the Finite Element Method in Biomedical Engineering, Biomechanics and Related Fields*, University of Ulm, 2009
29. L.O. Schwen, U. Wolfram, Validation of composite finite elements efficiently simulating elasticity of trabecular bone. Comput. Methods Biomech. Biomed. Eng. **17**(6), 652–660 (2014)
30. L.O. Schwen, U. Wolfram, H.J. Wilke, M. Rumpf, Determining effective elasticity parameters of microstructured materials, in *Proceedings of the 15th Workshop on the Finite Element Method in Biomedical Engineering, Biomechanics and Related Fields*, University of Ulm, July 2008, pp. 41–62
31. J. Stefan, Ueber die Theorie der Eisbildung, insbesondere über die Eisbildung im Polarmeere. Ann. Phys. **42**, 269–286 (1891)
32. T. Stein, Untersuchungen zur Dosimetrie der hochfrequenzstrominduzierten interstitiellen Thermotherapie in bipolarer Technik. Ecomed (2000). ISBN 3-609-20153-3
33. K. Stüben, A review of algebraic multigrid. J. Comput. Appl. Math. **128**(1–2), 281–309 (2001)
34. K. Ün, G. Bevill, T.M. Keaveny, The effects of side-artifacts on the elastic modulus of trabecular bone. J. Biomech. **39**, 1955–1963 (2006)
35. U. Wolfram, L.O. Schwen, U. Simon, M. Rumpf, H.J. Wilke, Statistical osteoporosis models using composite finite elements: a parameter study. J. Biomech. **42**(13), 2205–2209 (2009)
36. U. Wolfram, H.J. Wilke, P.K. Zysset, Rehydration of vertebral trabecular bone: Influences on its anisotropy, its stiffness and the indentation work with a view to age, gender and vertebral level. Bone **46**, 348–354 (2010)

Part II
Vascular System and the Brain

Chapter 12
Computational Simulation of Blood Flow and Drug Transportation in a Large Vasculature

Clément Coutey, Maxime Berg, Harvey Ho, and Peter Hunter

12.1 Introduction

Modelling the blood flow in a cardio-vascular system remains as a popular research topic due to its many clinical and physiological applications, and the complexity in biomechanical and mathematical methods to solve it. Through computer simulations we can investigate the blood flow in multiple spatial scales, from macro (cm) to micro (μm) levels, and in multiple temporal scales, from seconds to a much longer timeframe (e.g., months in case of chronic conditions), which are otherwise difficult to observe in laboratory experiments or animal models. This is particularly true if there are tiny vessels and control mechanisms involved. Indeed numerous models have been proposed over the last several decades, often coupled with in vivo or ex vivo experiments to validate the simulation results.

An investigation of the steady blood flow represents one of the modelling strategies since it removes temporal dependencies from the blood pressure and flow. In terms of computational cost, high efficiency can be achieved even for a large vasculature containing thousands or more vessels. Previous work in this area demonstrated real-time performance (in 50 ms) for a vasculature of 2337 vessels [1], whereas the works solving the pulsatile flow in the time domain (e.g., in [2]) and the frequency domain (e.g., in [3, 4]) rarely reached more than one thousand vessels.

We have several objectives in this study. Firstly we aim to implement a real-time blood flow solver similar to that of [1] and then apply it to large vasculatures containing up to thousands of vessels ranging from small arteries (diameter

C. Coutey • M. Berg
Department of Fluid Mechanics and Hydraulics, ENSEEIHT, Toulouse, France

H. Ho (✉) • P. Hunter
Auckland Bioengineering Institute, University of Auckland, Auckland, New Zealand
e-mail: harvey.ho@auckland.ac.nz

G. Joldes et al. (eds.), *Computational Biomechanics for Medicine*,
DOI 10.1007/978-3-319-28329-6_12

$D = \sim 1$ mm) to arterioles ($D = \sim 100\,\mu$m) and further to capillaries ($D = 10\,\mu$m). Secondly, we aim to simulate the transient flow of a drug or a contrast agent in that vasculature. Thirdly, we model the diffusion of the drug in a tiny tissue block of 1 mm, which can be used in future drug metabolism studies. All the algorithms have been developed using MATLAB (MathWorks, Natick, MA, USA) and the visualisation tool was CMGUI (http://www.cmiss.org/cmgui).

12.2 Method

12.2.1 A Real-Time Steady Flow Solver

The classic Hagen–Poiseuille law relates the pressure drop Δp to the flow rate q for a laminar, incompressible and Newtonian flow in a long cylindrical pipe as:

$$q = \frac{\Delta p}{R}, \tag{12.1}$$

where R represents the vessel resistance and is computed as:

$$R = \frac{8\mu L}{\pi r^4}, \tag{12.2}$$

where μ is the dynamic viscosity of the blood, L is the length of the vessel and r its radius. In order to solve the steady flow in a large tree, the flow at every branch is included in the flow rate vector Q of size M which is the number of branches, and the pressure p in the vector P of size N which is the number of nodes. Thus the Poiseuille law was computed through an $M \times N$ matrix K, so that $Q = KP$. Each line of K contains $R(i)$ which corresponds to the resistance of the branch i, and is multiplied by the pressure at the node ending the branch.

In order to solve the system, two boundary conditions need to be configured. Firstly the pressure at the beginning of the root vessel and the pressure at the end of terminal vessels. They were incorporated in the vector Pe and we created a matrix Γ to associate their values to the corresponding terms of P: $Pe = \Gamma P$. The second condition is similar to the Kirchhoff's first law which acts as an analog to blood flow. This means that the sum of inflow arriving at a node is equal to the sum of outflow leaving that node. The law was computed using a matrix ψ so that $\psi P = 0$. ψ is rather large due to the large number of internal nodes of the tree and each line contains $\sum \frac{\Delta P}{R(i)}$ corresponding to the sum of flow of all branches leaving that node. The final equation is:

$$\begin{bmatrix} Q \\ Pe \\ 0 \end{bmatrix} = \begin{bmatrix} K & \Gamma^T & \Psi^T \\ \Gamma & 0 & 0 \\ \Psi & 0 & 0 \end{bmatrix} \begin{bmatrix} P \\ \lambda_1 \\ \lambda_2 \end{bmatrix}, \tag{12.3}$$

where λ_1 and λ_2 are Lagrange multipliers resulting from the boundary conditions but treated as 1 in the actual implementation. In the first step the pressure P is evaluated from boundary conditions and the Kirchhoff law according to:

$$\begin{bmatrix} P \\ \lambda_1 \\ \lambda_2 \end{bmatrix} = \begin{bmatrix} K & \Gamma^T & \Psi^T \\ \Gamma & 0 & 0 \\ \Psi & 0 & 0 \end{bmatrix}^{-1} \begin{bmatrix} Q \\ Pe \\ 0 \end{bmatrix}. \tag{12.4}$$

The second step consists in using the matrix K in $Q = KP$. Name the first matrix at the RHS of Eq. (12.3) as H, the inversion of the matrix H in Eq. (12.4) is computationally expensive. However, since this step is performed before the actual computation of Q, it is treated as a pre-processing step and is not taken as the computational cost [1].

In order to check the computational time, we computed the steady flow in symmetric binary trees of an increased number of vessels. The number of branches in the tree was up to 8191 vessels and 14 generations, as shown in Fig. 12.1. The desktop computer used had an Intel Xeon CPU @2.67 GHz, and 4 GB of RAM. The computational time for these trees is shown in Table 12.1.

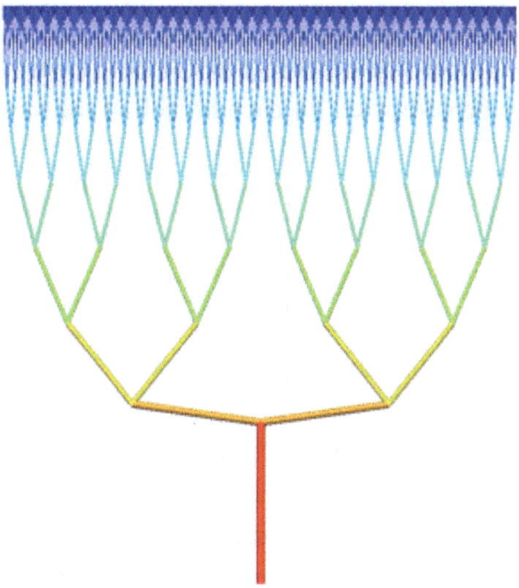

Fig. 12.1 Solve flow in a binary tree of 8192 vessels in real time (0.2 s)

Table 12.1 Computational time for arterial trees

Number of vessels ($n + 1$)	256	512	1024	2048	4096	8192
Time (ms)	0.5	1.6	5	16	56	206

12.2.2 A Transient Flow Solver

Assuming the blood flow is steady, the flow velocity u is evaluated by:

$$u = \frac{Q}{\pi r^2}. \tag{12.5}$$

The resulted u is then applied to the advection equation:

$$\frac{\partial C}{\partial t} + u\frac{\partial C}{\partial x} = 0, \tag{12.6}$$

where C is the concentration of a drug agent. The finite difference method was used to solve the equation, which in turn requires a 1D mesh created for each vessel. In order to reduce numerical diffusion the spatial step needs to be sufficiently small albeit the computational time is higher. We used an implicit-scheme which is an unconditionally stable scheme so that the constraints on the time step can be loosened. The final scheme is:

$$C_i^n = C_i^{n+1}(1+\alpha) - \alpha C_{i-1}^{n+1}, \tag{12.7}$$

where i denotes the position in space and n in time. $\alpha = \frac{u\Delta t}{\Delta x}$ is the Courant number. With a different vessel length, Δx can be different and therefore α has to be computed for each branch of the vasculature. Finally we have:

$$\begin{bmatrix} \vdots \\ \\ C_i^n \\ \\ \vdots \end{bmatrix} = \begin{bmatrix} 1+\alpha & & & \\ -\alpha & & & \\ & \cdots & & \\ & & \cdots & \\ & & -\alpha & 1+\alpha \end{bmatrix} \begin{bmatrix} \vdots \\ \\ C_i^{n+1} \\ \\ \vdots \end{bmatrix}. \tag{12.8}$$

As for the initial conditions, since a contrast agent is injected at the root of the tree, and so a C_0 can be configured for the root vessel. At bifurcations, the concentration at the start (or proximal end) of the daughter branch is $C_d = C_m \frac{r_d^2}{2r_m^2}$ where the subscript m denotes mother vessel and d denotes the daughter vessel.

12.2.3 CCO Tree Growing Algorithm

Assuming that a vasculature grows according to the principle of energy minimisation, i.e., the vasculature uses the minimum energy to perfuse a tissue, an optimisation process namely Constraint Constructive Optimisation (CCO) can be used to generate the tree [5]. The core of this method is the minimisation of a target function, which is the total blood volume as [5] suggested. Mathematical details of the algorithm can be found in [5, 6]. In brief, the tree generation algorithm is summarised as below:

1. Randomly generate a point within the perfusion volume;
2. Search for the closest segment to the point;
3. Perform an optimisation operation to create a bifurcation;
4. Check if all constraints are satisfied;
5. Generate a list of candidates which passed the test of Step 4 and
6. Use the one candidate which has the smallest tree volume.

With this algorithm we created arterial trees of different number of vessels (100, 200, 2000 and 4000) as shown in Fig. 12.2. The steady and transient flow solvers described in the above two sections were then applied to the trees.

12.2.4 Tree Growing Algorithm Under the Fahraeus–Lindqvist Effect

When the blood reaches the scale of arteriole (from 8 to 100 μm), in order to minimise the amount of energy required to move the blood, red blood cells spontaneously migrate toward the centre of the vessel. This increases the shear rate and lowers the viscosity. Since the diameter of red blood cells is comparable to the diameter of arteriole, the blood viscosity becomes dependent on the vessel radius as well. This phenomenon is called Fahraeus–Lindqvist effect, which was firstly reported in 1931 (for a recent account of this effect please refer to [7]).

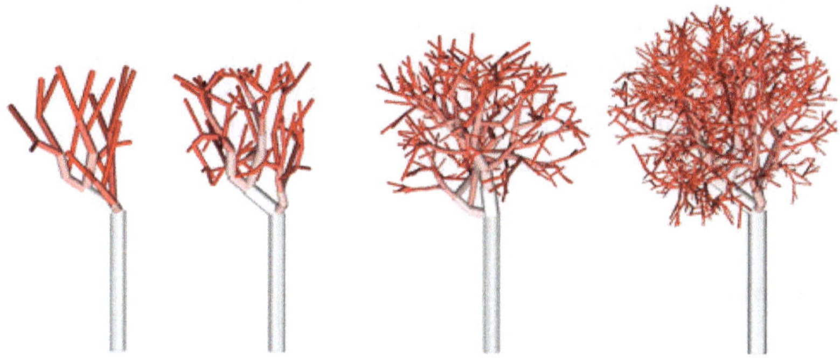

Fig. 12.2 Arterial trees generated from the CCO algorithm. The number of vessels in the tree is 100, 200, 2000 and 4000, respectively

The Fahraeus–Lindqvist effect was also implemented in the vessel tree growing algorithm. Specifically the following rule according to [6] was enforced:

$$\mu(r) = \frac{\mu_\infty}{(1 + \delta/r)^2} \tag{12.9}$$

with $\mu_\infty = 4.0\,\mathrm{cP}$, $\delta = 4.29\,\mu\mathrm{m}$ in the range $r \in [4150]\,\mu\mathrm{m}$. Therefore this rule relates the blood viscosity with the vessel radius.

12.2.5 The 3D Diffusion Solver

While the tree generation algorithm can be applied to both macro- and micro-scales, of particular interest is the tree generated for a tiny tissue block of 1 mm in size, where the terminal vessels reached the μm scale, i.e., the size of capillaries. At this spatial level the exchange of nutrients, oxygen and metabolism by-products between the vascular network and the interstitial space needs to be considered, where diffusion is the major transportation mechanism. The transient diffusion equation in 3D can be expressed as [8, 9]:

$$\frac{\partial c(r,t)}{\partial t} = D \cdot \left(\frac{\partial^2 c(x,y,z,t)}{\partial x^2} + \frac{\partial^2 c(x,y,z,t)}{\partial y^2} + \frac{\partial^2 c(x,y,z,t)}{\partial z^2} \right) - k(x,y,z), \tag{12.10}$$

where $D = 1.0 \times 10^{-9}\,\mathrm{m^2\,s^{-1}}$ is the diffusion coefficient for oxygen [8], $c(\mathbf{r}, t)$ is the transient concentration in the whole domain at location $\mathbf{r}$ and time t. k is the consumption term describing drug uptake by cells in the tissue block.

In order to solve Eq. (12.10), an explicit Euler scheme was used whereby the spatial step was fixed (10 μm), and the temporal step was chosen to obtain a stable scheme for diffusion. Stability criterions are given by $\beta = \frac{D * \Delta t}{\Delta x^2} < 0.5$, which gives us $\Delta t < 5.0 \times 10^{-2}$ s and therefore $\Delta t = 1$ ms was chosen. The computational time for a 3D grid of $100 \times 100 \times 100$ (grid size 10 μm) was 450 s for 1000 iterations.

More details of the numeric scheme were introduced in [9] and we refer the interested reader to that literature for reference.

12.3 Results

12.3.1 Real-Time Steady Flow Solving for an Arterial Tree

The steady flow solver was applied to arterial trees generated from the CCO algorithm. The computations were run in real time, as described in Table 12.1. Figure 12.3 visualises the flow rate (ml/s) distribution in a tree of 512 branches.

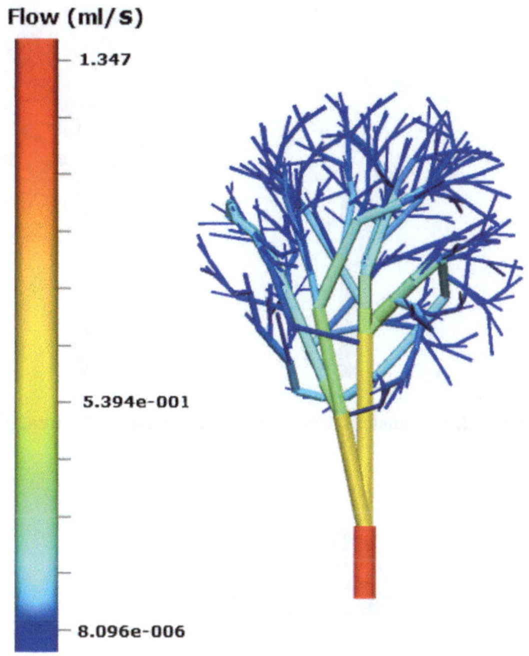

Fig. 12.3 Real-time flow solving in an arterial tree generated from the CCO algorithm

The blood vessel radii in the tree range from 1 mm at the root vessel to 10 μm in terminal vessels.

Due to the law of mass conservation, the flow rate at each generation was identical to other generations. Therefore the flow rate gradually decreases as the flow diverged in downstream generations. At the root vessel, the flow rate is 1.35 ml/s. Considering the radius r of the root vessel is 1 mm, translated into flow velocity this is equivalent to 44.6 cm/s.

12.3.2 Drug Convection in the Tree

We assume that there was no leakage occurring in the tree before the terminal capillary vessels. The flow velocities computed from the previous section were used to simulate the evolution of a drug agent through the tree by setting u values in Eq. (12.6).

In the finite difference scheme introduced in Sect. 12.2.2, each vessel in the tree was split into 4000 nodes to solve the convection equation. Consequently the computational time was much longer (~1000 s for 512 vessels) than the steady flow solver. For a better visualisation of the agent distribution in the tree, a logarithmic scale was used, as shown in Fig. 12.4. The concentration of the agent became lower while it progressed in the tree. This is understandable because the total section area of the vasculature is increasing.

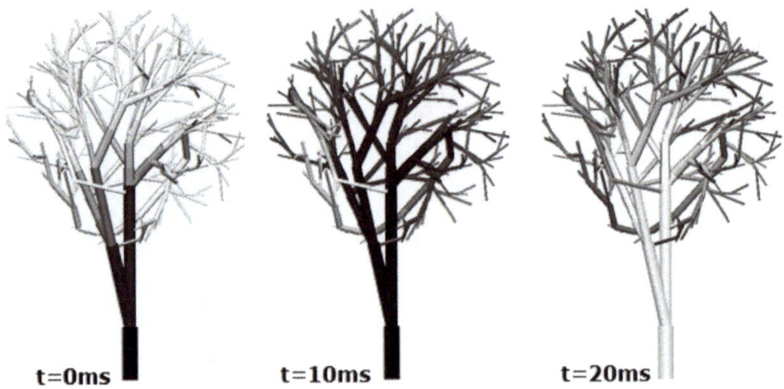

Fig. 12.4 Transient flow of a contrast agent in the tree. No leakage was assumed before terminals of the tree

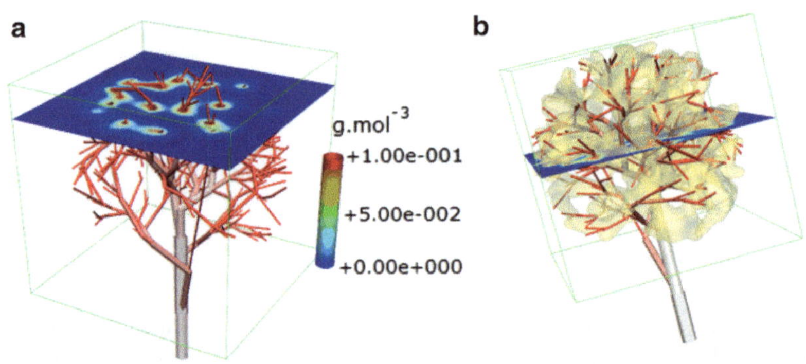

Fig. 12.5 Concentration of a drug agent in a tissue block due to diffusion effects from the micro-vasculature. (**a**) Visualisation from a cut plane; (**b**) visualisation from the evolution of the isosurface of drug concentration

12.3.3 Diffusion from the Tree

Figure 12.5 shows drug concentration C in the tissue as a result of diffusion from the micro-scale vasculature. In Fig. 12.5a a cut plane was used to show C at a cross-section of the tissue block. It can be seen that C decreases while the distance from the vessels increases. Figure 12.5b shows an isosurface where the drug concentration was identical on the surface.

12.4 Discussion and Conclusion

In this paper we presented a multiscale computational pipeline which contains a real-time steady flow solver, a transient agent flow solver and a diffusion solver. This pipeline was applied to a tree generated from the CCO algorithm and different flow phenomena were simulated.

One of the applications of this pipeline is to model the hepatic vasculature, where the large liver vessels (portal and hepatic veins) are digitised from CT/MRI images, and the CCO algorithm can be used to generate small (<1 mm) vessels downstream the large vessels. The steady flow solver, introduced in Sect. 12.2.1, is suitable for the venous flow as there are little blood pressure fluctuations in these veins. The convection and diffusion algorithms may be applied to the study of nutrient metabolism and drug detoxification, after incorporating some recently published models (e.g., that for acetaminophen hepatotoxicity [10]).

It worths mentioning that the geometry of the arterial tree, including the nodes and elements, was organised into a 1D finite element mesh. This approach is different from the graph method described in [11]. When embedded in a 3D finite element volume mesh the relative coordinates (or ξ coordinates) of the vasculature can be evaluated dynamically and deformed in real time. This concept is also called *host-mesh fitting* and has been described in other relevant studies, e.g., for musculoskeletal modelling [12]. Its application for the vascular network, illustrated in Fig. 12.6, is powerful in surgical simulations or computer aided surgeries where organs are deformed due to surgical loads or breathing effects.

There are some limitations pertaining to the current pipeline. Firstly the diffusion process introduced in Sect. 12.2.3 was rather idealistic. For example, the k term in Eq. (12.10) which accounts for the cellular uptake is artificial but the actual perfusion process across cellular membranes is very complex. Also the tissue was modelled as a homogeneous media which indeed should have heterogeneous resistance to oxygen perfusion. Secondly the blood was simulated as Newtonian, i.e. with a constant viscosity which indeed should be a variable corresponding to the shear rate, i.e. as a shear thinning non-Newtonian fluid. This assumption becomes more questionable in tiny vessels where the diameter of red blood cells is close to the vessel lumen. Thirdly the coupling between the different solvers is not implemented yet, in particular between the extra-cellular and intra-cellular models.

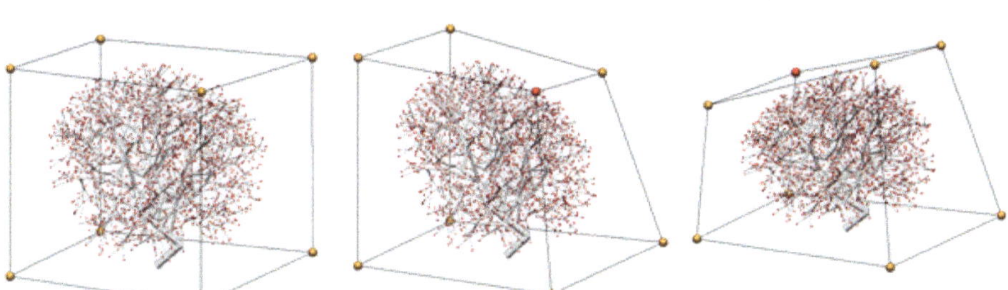

Fig. 12.6 Real-time vasculature deformation based on the host-mesh fitting algorithm

Nevertheless, the presented computational pipeline may be used in some biomedical applications due to its computational efficiency and also the ability to model drug transportation. Future work includes the investigation of cellular reactions and metabolism functions, and strong coupling of the solvers.

Acknowledgements We thank Alexandre Muller and Alice Chapuis for their previous work in the CCO algorithm and the 3D diffusion work.

References

1. X. Wu, J. Allard, S. Cotin, Real-time modeling of vascular flow for angiography simulation, in *Proceedings of the 10th International Conference on Medical Image Computing and Computer-Assisted Intervention - Volume Part I* (Springer, Berlin, 2007), pp. 557–565
2. P.J. Blanco, S.A. Urquiza, R.A. Feijóo, Assessing the influence of heart rate in local hemodynamics through coupled 3d-1d-0d models. Int. J. Numer. Methods Biomed. Eng. **26**(7), 890–903 (2010)
3. A.P. Avolio, Multi-branched model of the human arterial system. Med. Biol. Eng. Comput. **18**, 709–718 (1980)
4. D.P. Segers, N. Stergiopulos, P. Verdonck, R. Verhoeven, Assessment of distributed arterial network models. Med. Biol. Eng. Comput. **35**(6), 729–736 (1997)
5. W. Schreiner, P. Buxbaum, Computer-optimization of vascular trees. IEEE Trans. Biom. Eng. **40**(5), 482–491 (1993)
6. L.O. Schwen, T. Preusser, Analysis and algorithmic generation of hepatic vascular systems. Int. J. Hepatol. 1–17 (2012)
7. A.R. Pries, T.W. Secomb, Microvascular blood viscosity in vivo and the endothelial surface layer. Am. J. Physiol. Heart Circ. Physiol. **289**(6), H2657–H2664 (2005)
8. S.M.B. Shrestha, G.R. Joldes, A. Wittek, K. Miller, Cellular automata coupled with steady-state nutrient solution permit simulation of large-scale growth of tumours. Int. J. Numer. Methods Biomed. Eng. **29**(4), 542–559 (2013)
9. A. Chapuis, H. Ho, A computer simulation for 3d vasculature-based oxygen distribution and tumour growth, in *Computational Biomechanics for Medicine*, ed. by B. Doyle, K. Miller, A. Wittek, P.M.F. Nielsen (Springer International Publishing, Cham, 2015), pp. 25–35
10. R. Ben-Shachar, Y. Chen, S. Luo, C. Hartman, M. Reed, H.F. Nijhout, The biochemistry of acetaminophen hepatotoxicity and rescue: a mathematical model. Theor. Biol. Med. Model. **9**, 55 (2012)
11. D. Selle, B. Preim, A. Schenk, H.O. Peitgen, Analysis of vasculature for liver surgical planning. IEEE Trans. Med. Imaging **21**(11), 1344–1357 (2002)
12. K. Oberhofer, K. Mithraratne, N.S. Stott, I.A. Anderson, Anatomically-based musculoskeletal modeling: prediction and validation of muscle deformation during walking. Vis. Comput. **25**(9), 843–851 (2009)

Chapter 13
Fundus Image Based Blood Flow Simulation of the Retinal Arteries

Andreas Kristen, Lachlan Kelsey, Erich Wintermantel, and Barry Doyle

13.1 Introduction

The retina is affected by numerous systemic diseases related to the vascular circulation, like diabetes, hypertension, and atherosclerosis, which are major health problems in modern society [8]. Several studies have shown that narrowing of retinal arterioles and widening of the retinal veins indicate cardiovascular health risk including coronary heart disease (CHD), hypertension, and risk of stroke, even before clinical symptoms of the disease occur [13, 27]. Therefore, the eyes represent a window into the cardiovascular health of a person. The retinal vasculature has blood vessels with diameters less than 150 μm and includes the small resistance arteries, arterioles, capillaries, and venules in the microcirculation [26]. These make up the largest part of the circulation system of a person, but it is still not possible to examine this system non-invasively, except in the retina. Compared to large arteries, vessels of the microcirculation have different rheological properties due to the physiological and physical limit [6].

Therefore, it is necessary to understand the hemodynamics of the retinal vasculature and investigate the microcirculation system. Fundus photography is a widespread and easy to perform imaging method to gain a view of the retinal blood

A. Kristen • L. Kelsey • B. Doyle (✉)
Vascular Engineering, School of Mechanical and Chemical Engineering,
The University of Western Australia, Perth, WA, Australia
e-mail: andreas.kristen@uwa.edu.au; lachlan.kelsey@student.uwa.edu.au;
barry.doyle@uwa.edu.au

E. Wintermantel
Institute of Medical and Polymer Engineering, Faculty of Mechanical Engineering,
University of Technology Munich, Garching, Germany
e-mail: erich.wintermantel@tum.de

G. Joldes et al. (eds.), *Computational Biomechanics for Medicine*,
DOI 10.1007/978-3-319-28329-6_13

vessel network, consequently, the microcirculation of a person. A realistic model of the retinal vascular network can improve the knowledge of the hemodynamics and may help physicians to detect abnormalities earlier.

The aim of this work is to investigate the blood flow in the retina by computational modelling and a computational fluid dynamics (CFD) simulation. The data for the model are obtained from a fundus photograph. In all visible branches of the plane 2D retinal artery network the velocity and pressure distribution is examined.

13.2 Methods

For this work we used a high-resolution fundus image from the free High-Resolution Fundus (HRF) Image Database (Friedrich-Alexander Universitaet Erlangen-Nuernberg) [3] of a healthy subject. The image of the retina were taken by an expert with a CANON CF-60UVi camera and has a resolution of 3504×2336 pixels. The vessels were segmented with the method described by Budai et al. [4] using the Frangi algorithm [32] for ridge detection. Frangi filter method extracts tubular objects based on measurements of the eigenvalues of the Hessian matrix. The Hessian matrix contains the second-order derivatives in a local neighborhood. Before applying the Frangi filter, the RGB image was decomposed into the green channel, because this offers the best illumination [4]. Then histogram stretching and homomorphic filtering were performed to normalize and increase the contrast as well as the brightness and remove noise. The Frangi algorithm was applied on the image with $\sigma = 1, 2 \ldots 8$, on a rescaled image by the factor of 0.5 with $\sigma = 1, 2, 3, 4$ and on a rescaled image by the factor of 0.25 with $\sigma = 1, 2$. The values were obtained from Frangi et al. [32] with empirical gained adaption to get the best fitting results in relation of vessel detection and noise. σ is the standard deviation of Gaussian to approximate the second-order derivatives. The filter correction constants β_1 and β_2 were set to 2 and 4. The images are converted to binary images by a threshold and rescaled back to the original size. The final segmented blood vessels are the result of the original and rescaled superimposed images. An increase in the value of σ increases the thickness of the segmented vessels. Therefore, we used the superimposition of the original image to guide our segmentation process to ensure the vessels remained true to the original geometry. The original fundus image with the green channel and the segmented vessels are shown in Fig. 13.1.

Since the CFD simulation is performed on the artery tree, the segmented vessels have to be divided into arteries and veins. There are reports in the literature and active research trying to automatically classify arteries and veins, like from Kondermann et al. [12], but this is not the focus of this study. The classification was done manually and the arteries were marked on the original image with red lines and the veins with blue lines. A special written MATLAB (MathWorks, USA) script compared the original image with the added lines and the segmented blood

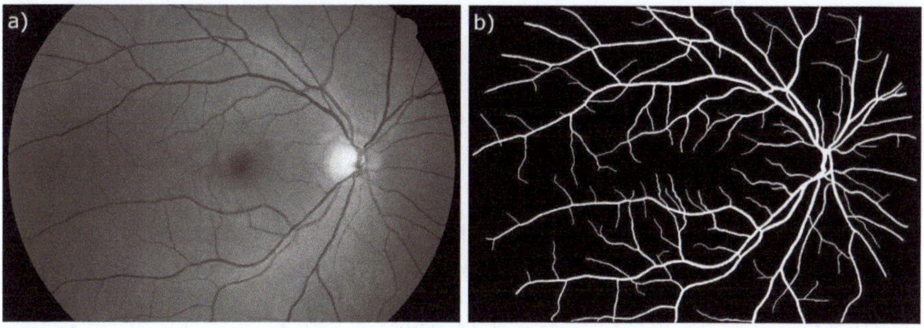

Fig. 13.1 (**a**) The basis fundus image of a healthy subject taken from the High-Resolution Fundus (HRF) Image Database [3] extracted to the green channel. (**b**) Segmented blood vessels after applying Frangi filter algorithm

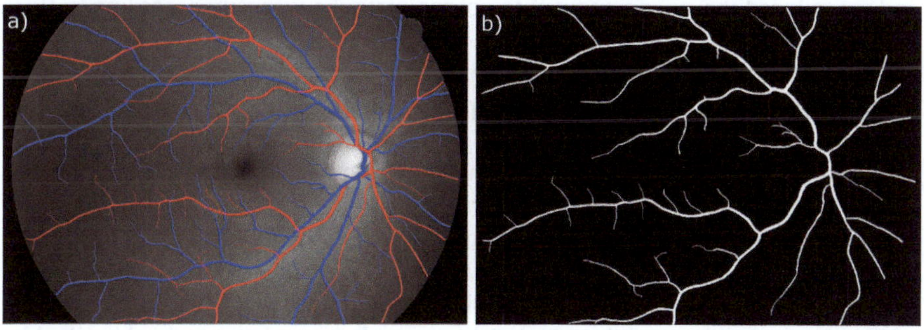

Fig. 13.2 (**a**) Classification of the retinal blood vessel network into arteries marked *red* and veins marked *blue*. (**b**) Arteries divided from the segmented blood vessels

vessels to generate a new binary image containing just the arteries or veins. The classification into arteries and veins as well as the segmentation showing only the artery network is displayed in Fig. 13.2.

With Mimics v18 (Materialise, Belgium) a mask was created of the binary file shown in Fig. 13.2b, which was scaled into mm by using the standard optic disc diameter assumed to be 1.85 mm [10, 11] and exported into 3-matic v10 (Materialise, Belgium) afterwards. A smoothed geometry curve was generated to export the segmented blood vessels as an STL or CAD file containing a plane surface. The geometry could then be imported directly into ANSYS Workbench v15 (Ansys, USA) to mesh the structure, define the boundary conditions, and run the CFD simulation with ANSYS Fluent v15.

For the mesh of the vessel structure, quadrilateral elements were used including on each wall boundary inflation with 20 layers starting on the wall with a size of 0.5 μm and growing with a factor of 1.2 into the center [1]. The mesh in the center contains elements with a constant size of 2 μm. The final mesh had 2,551,978 elements. Figure 13.4b shows an extract of the mesh.

The blood flow in the retinal arteries is assumed to be steady and governed by the Navier–Stokes equation for incompressible fluid [2, 14, 21]

$$\nabla \cdot \mathbf{V} = 0$$
$$\varrho(\mathbf{V} \cdot \nabla)\mathbf{V} = -\nabla p + \mu \nabla^2 \mathbf{V} \tag{13.1}$$

where V is the velocity of the blood, p the pressure, ϱ the blood density, and μ is the dynamic viscosity of the blood. The vessel walls were assumed to be rigid [14] and the density of the blood was set to 1055 kg/m^3 [20]. Because the diameters of the visible arteries are in the range of 15–120 μm and the complex characteristic of blood, a non-Newtonian fluid model has to be considered. In our simulation we used the Carreau–Yasuda non-Newtonian fluid model where $\eta_0 = 2.5\,\text{mPa s}$, $\eta_\infty = 160\,\text{mPa s}$, $\lambda = 8.2\,\text{s}$, $n = 0.2128$, and $a = 1,23$ adapted from Cho et al. [5] and we reduced the η_∞ component to approximate the viscosity according to Pries et al [23].

As some of the smaller vessels are not visible on the image or were not detected by the segmentation, the unresolved peripheral vessels were modelled by generating structural fractal trees [19, 28]. The same was applied for the vessels ending abruptly on the edge of the fundus image due to the limitation of 2D planar photographs. The outlet diameters in the image are in the range of 14–86 μm. All vessels terminating with diameters larger or equal to 30 μm are connected to an asymmetric binary structured tree, where at each bifurcation the radius of the two daughter vessels is scaled by factors α and β. The relationship across bifurcations between the radius of the parent vessel r_p and the two daughter vessels r_{d1} and r_{d2} can be described by the power law

$$r_p = {r_{d1}}^{\xi} + {r_{d2}}^{\xi} \tag{13.2}$$

and the asymmetry index γ

$$\gamma = \frac{{r_{d2}}^2}{{r_{d1}}^2} \tag{13.3}$$

where ξ, the junction exponent, set to 3 [18] and γ to 0.62 and 0.41, respectively, depending on the vessel radius [16, 33]. A schematic structured fractal tree can be viewed in Fig. 13.3.

To generate the vascular bed, the branches of a fractal tree terminate, when the daughter vessel reaches a diameter below 30 μm and has an equal relative pressure P_{end} of 0 mmHg [14]. The outflow boundary condition at each outlet is given by the pressure drop

$$\Delta P = P_0 - P_{\text{end}} = R_0 \times Q_0 \tag{13.4}$$

where R_0 is the total resistance of each fractal tree, Q_0 the volumetric outflow rate, and P_0 the outflow pressure at the outlet of the trunk arteries [14, 20]. The total resistance of each fractal tree is calculated iteratively in recursive manner starting from the terminal branch.

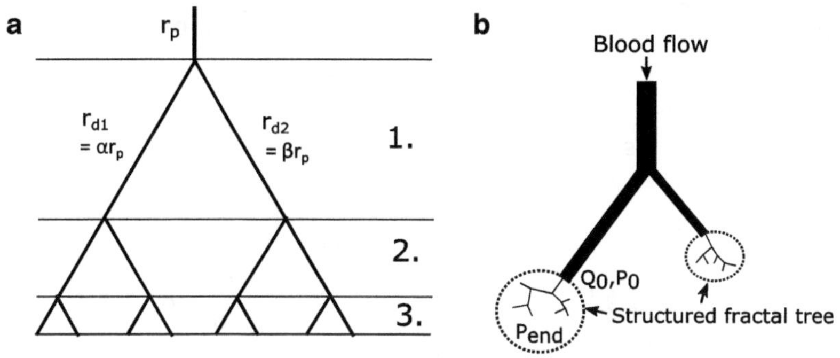

Fig. 13.3 (**a**) An asymmetric binary fractal tree. (**b**) Schematic structural fractal tree connected to the trunk vessel

A constant velocity from the central retinal artery of 0.07 m/s [17] was assumed as the inlet flow. The no-slip condition was used for the wall boundaries, where the wall shear stress (WSS) at the wall interface is infinite and thus the velocity is zero [7]. The Navier–Stokes equation was solved numerically by the commercial finite volume solver ANSYS Fluent.

The visible arterial network has two inlets, where the central retinal artery (CRA) enters the network in the optic disc and branches into the superior temporal and inferior temporal arteriole [30]. The outflow occurs via 52 outlets (see Fig. 13.4). To take account that the inlet flow of the CRA is divided into two arteries the velocity-inlet boundary is set by the volume flow rate $v_{CRA} \times A_{CRA} = v_1 \times A_1 + v_2 \times A_2$ with $v_1 = v_2$ and $A_1 = A_2$. Resulting at each inlet a velocity of 0.037 m/s by using a mean CRA diameter of 175 μm [9].

13.3 Results

The computing time of the simulation was 1.5 h until the solution converged to residual values of 10^{-3} for continuity and x and y velocity. We performed the simulation using an Intel(R) Core(TM) i7 960 CPU 3.20 GHz personal computer with 12 GB RAM. Figure 13.5 shows the blood flow velocity distribution in the retinal arterial network. The velocity vectors at three bifurcations (junction J_1, J_2, and J_3) are displayed in Fig. 13.6. At each bifurcation the flow rate as well as the velocity of the blood is reduced. At the ending branches after several bifurcations the blood flow reaches a nearly zero velocity (1×10^{-5} m/s). Exceptions are vessels with less bifurcations leaving the visible vascular network near the optic disc (e.g., O_2, O_{30}). The velocity profiles were close to parabolic and became flatter downstream (see Fig. 13.7).

Figure 13.8 shows the pressure distribution of the retinal arterial network. The pressure drops further away from the optic disc to the terminating branches to

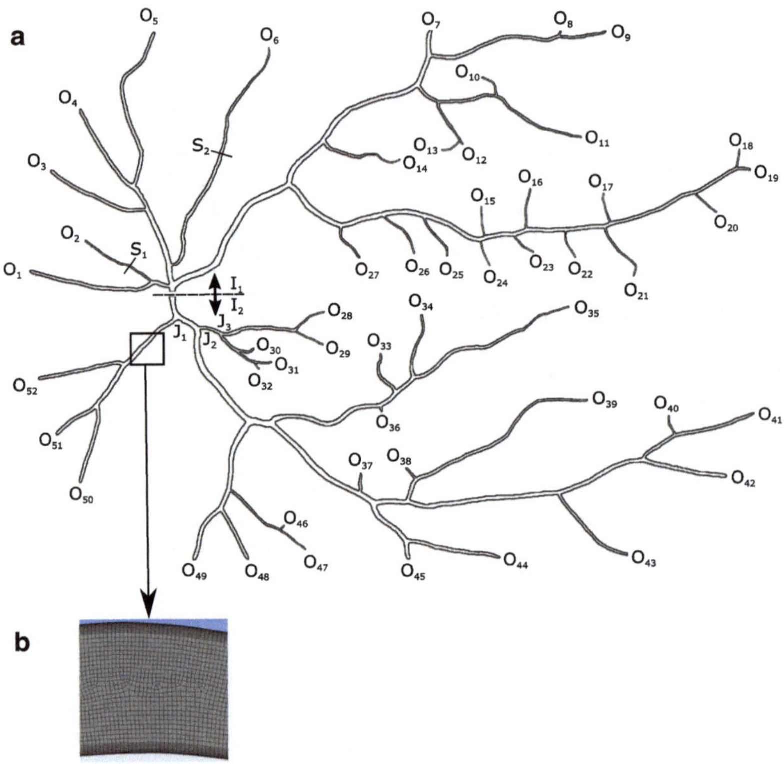

Fig. 13.4 (**a**) Retinal arterial vessel network with 52 outlets represented by "O_i" and a split at the optic disc for two inlets represented by "I_i." (Rotated through 180° compared to the raster image.) (**b**) Extract of the mesh

around 14–15 mmHg, whereas the vessels leaving the visible vascular network near the optic disc maintain a high pressure level. In Table 13.1 we show the pressure drops and velocities for outlets with large diameters to small and outlets near the optic disc to peripheral.

13.4 Discussion

The aim of this study was to create an initial step towards comprehensive patient-specific models of the retinal vasculature based on readily available imaging. We have investigated the blood flow of the retinal artery network by computational fluid dynamics using geometry extracted from a high-resolution fundus image.

In previous studies in fundus image based CFD simulations from Liu et. al [14] and Malek et al. [15] their arterial network had ten and nine outlets, respectively, compared to 52 outlets in our study. Therefore, their vessel network represents

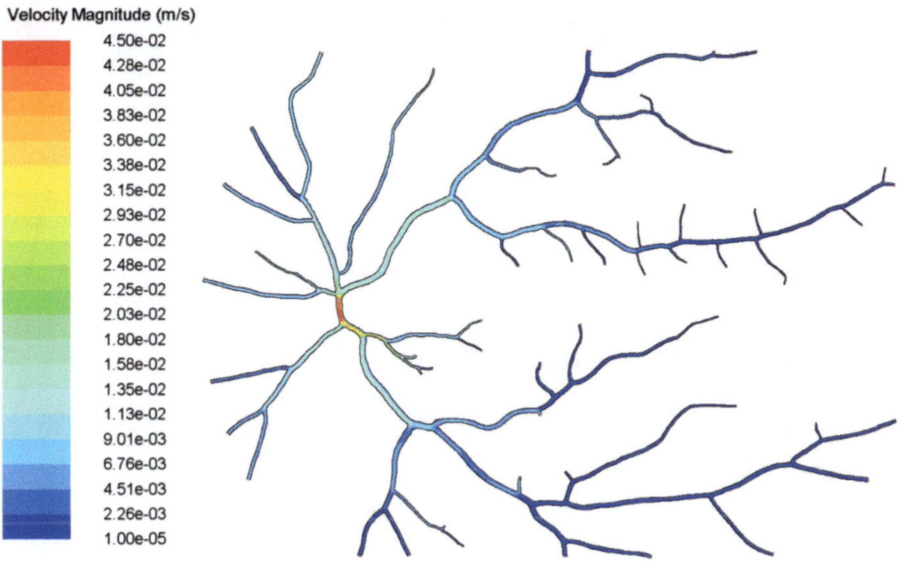

Fig. 13.5 Contour plot of the velocity distribution of the retinal arterial network

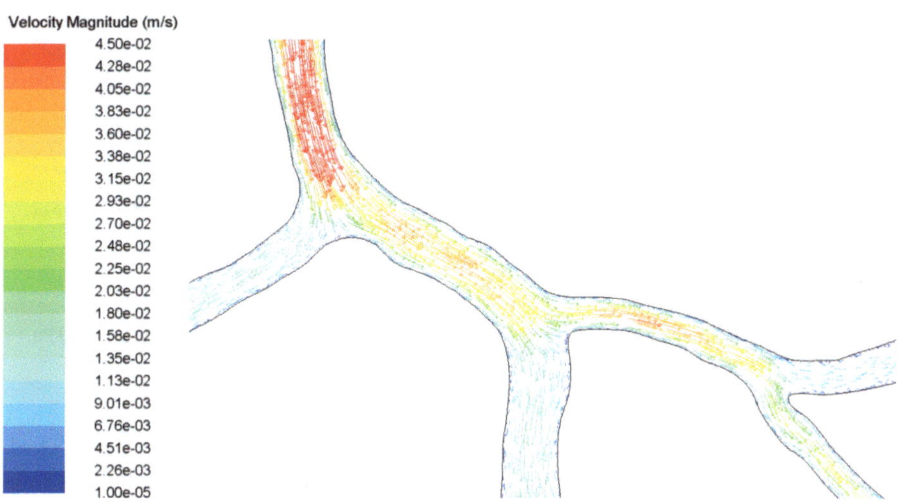

Fig. 13.6 Vector plot of the velocity at junction J_1, J_2, and J_3

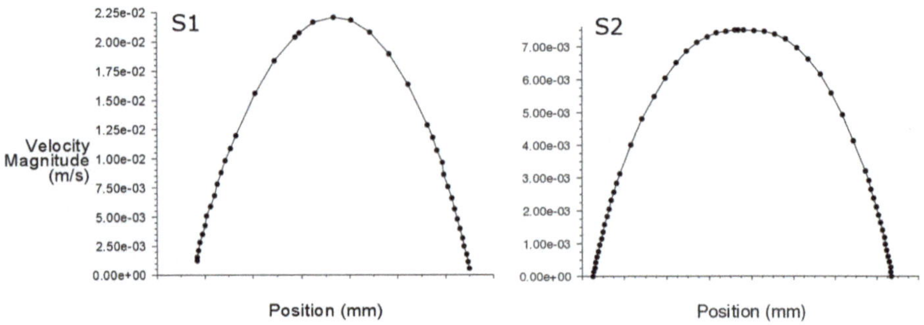

Fig. 13.7 Velocity profiles at section S_1 and S_2

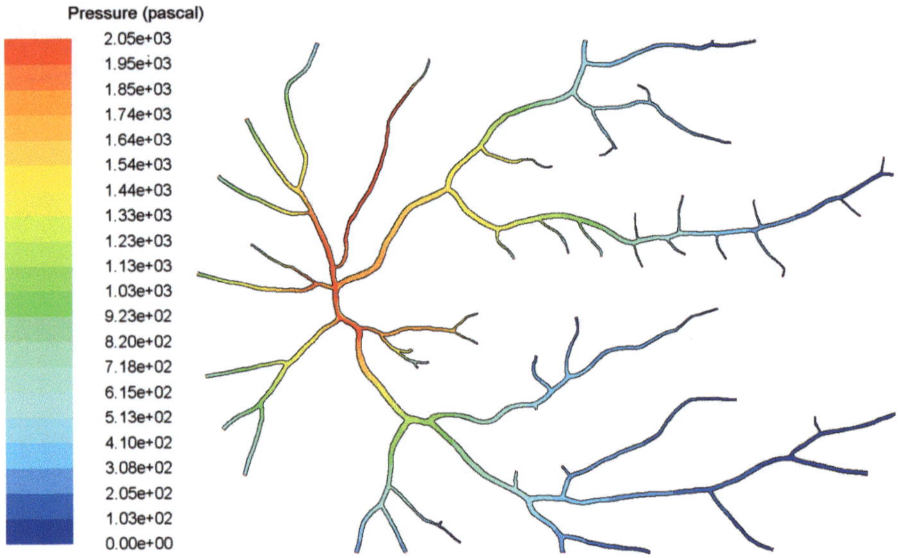

Fig. 13.8 Contour plot of the pressure distribution of the retinal arterial network (1000 Pa ≈ 7.5 mmHg)

a simplified geometry of the retinal arteries. In Liu et al. [14] and Malek et al. [15] the inlet flow velocity in their simulation was set equal to the velocity in the central retinal artery, which may be inappropriate, because the CRA branches when entering the retina in the optic disc. This is the reason why our velocity values are approximately half of their results. For instance, if we compare the velocity at locations S_1 and S_2 (see Fig. 13.4) with the data from Malek et al. [15] we notice that the velocity at S_1 peaks at 0.025 m/s (compare to 0.06 m/s) and at 0.0075 m/s at S_2 (compare to 0.015 m/s). The parabolic form of the velocity profiles is similar to the results from Liu et al. [14] and Malek et al. [15].

At each bifurcation the flow rate as well as the velocity of the blood is reduced, which shows the calculated velocity distribution correctly. This is also in correlation

Table 13.1 Calculated pressure drop and mean velocity at outlet

Location	Diameter (μm)	Pressure drop (mmHg)	Mean velocity (cm/s)
O_1	77	8.55	0.93
O_2	39	12.38	2.31
O_3	83	10.08	0.84
O_4	74	9.47	0.52
O_5	75	12.75	1.02
O_6	26	15.38	2.25
O_{20}	30	15.38	0.01
O_{21}	34	15.38	0.04
O_{29}	30	13.22	0.43
O_{30}	50	7.80	1.80
O_{31}	34	13.76	0.07
O_{40}	24	15.38	0.03
O_{41}	49	15.36	0.004
O_{43}	60	15.27	0.02
O_{44}	55	15.20	0.34
O_{45}	86	13.91	0.23

with blood flow velocity measurements with bidirectional laser Doppler velocimetry and calculated volume flow rates in the human retinal arteries performed by Riva et al. [25].

Yet, there are still no direct measurements of pressure drops in human retinal arteries [14], but Quigley et al. [24] estimate a pressure drop from the optical disc to the peripheral vessels with diameters of 30–40 μm to be around 15 mmHg, which correlates well with our study. Also the pressure distribution shows similarities with the results from Malek et al. [15], except that in our study the pressure around the optic disc is higher (≈15.4 mmHg (2050 Pa) compared to 14.8 mmHg). It is not possible to compare vessels leaving the network near the optic disc as in Malek et al. [15] they did not have any vessels with outlets near the optic disc.

Due to the aim of this study, that is, to use cheap and readily available imaging methods, our geometry is limited to planar 2D. Thus, the 3D geometry, like the curvature of the retina and the tubular form of the vessels, is ignored. Due to the sphere geometry of the retina, the length and diameters of the vessel branches are not represented correctly in a plane 2D model. Furthermore, the volumetric flow rate behaves different in a 2D simulation and consequently the pressure (refer to $\Delta P = R_0 \times Q_0$), because the cross section is calculated as a rectangle and not with $r^2 \times \pi$. Thus, the results related to the flow rate in the present study have to be taken as qualitative values instead of quantitative. Using optical coherence tomography (OCT) as the imaging input, 3D information would be available including blood vessels down to the capillary bed. However, using OCT to image retinal vasculature presents many challenges. Firstly, when the OCT device does not have a tracking system, it is only possible to image a field of view up to 6 × 6 mm. For better quality images, ophthalmologists usually scan even smaller areas. Furthermore, motion artifact and phase-noise are visible in these images. Although, fundus images are limited to 2D, they offer clear visible blood vessels and in high resolution. A fundus image is limited to a field of view, which cuts the vessels on its edge abruptly. This is a major problem within large vessels, which are ending with diameters of 70–85 μm. Wide angle cameras with a large field of view could be a solution, but these images would lose spatial resolution.

The assumption of a steady blood flow does not consider cardiac cycles. Wang et al. [31] performed in vivo retinal blood flow measurements in the vessels entering and leaving the optic disc by Fourier domain Doppler OCT. Furthermore, the pulsatile flow was measured and averaged over one cardiac cycle resulting in a total arterial flow of 53.89 μl/min [31], which is similar to the total blood flow in our arterial system with 53.57 μl/min ($Q_{\mathrm{In1}} + Q_{\mathrm{In2}}$), even with the 2D geometry. In further studies, we plan on verifying this using transient simulations.

For this work we did not perform a mesh independence study, however, due to the number of 2,551,978 elements, we assume that the mesh is refined enough.

In our simulation we used the Carreau–Yasuda non-Newtonian fluid model, which is used in general blood flow simulations, but it does not consider the Fahraeus–Lindqvist effect in microcirculation, where the viscosity depends on the hematocrit and vessel diameter [23] and increases in vessels with diameters below 40 μm [22] (hematocrit $\mathrm{H_D} = 0.45$). However, in the diameter range of 40–120 μm, which constitutes 70 % of our vessels, with shear rates of 1500–1800 1/s [29], both models provide nearly similar dynamic viscosities. In a mathematical model about hemodynamic parameters in the human retina vasculature [29], the apparent viscosity decreases even until a vessel diameter of 7.4 μm, which covers our complete vessel network and thus the approximation of the Carreau–Yasuda model should be valid.

The invisible vessels or ending vessels due to the limitation of planar 2D photographs are represented by a structural tree, which makes it possible to generate the outlet boundaries, but it does not adequately represent the real branching of the vessels and their physical parameters. Further work is needed to perfect the vessel segmentation method and high-resolution images to detect all vessels down to the vascular bed of 4 μm. Until then, the structural fractal tree is one of the few reliable methods to model downstream resistance. In this study we used a terminating criteria, like the previous blood flow simulation studies of Liu et al. [14] and Malek et al. [15], of a vessel diameter below 30 μm, where the pressure is set to $P_{\mathrm{end}} = 0$. Sixteen of the 52 outlets have diameters below 30 μm and are assumed as the final vessel bed, which generates an abrupt pressure change inside the vessel system. As a consequence, the blood flow accelerates in front of the outlet and creates unphysiological conditions (see outlets O_6, O_{14}, and O_{26}). The attempt to calculate down to 14 μm as terminating criteria failed due to unrealistic results. In further studies the structural fractal tree must be designed down to vessel diameters below the smallest vessel in the model. Takahashi et al. [29] stated in their study that the pressure drop of the vascular bed after the arterioles to the capillaries occurs only within a vessel diameter of 5 μm and before the intravascular pressure in the retinal arteries decreases from 38.9 mmHg at a vessel diameter of 108 μm to 29.3 mmHg at a vessel diameter of 5.1 μm.

Further plans for our study are to connect this retinal artery network to the venous network and create a 3D model of the retinal vasculature.

13.5 Conclusion

In this study we presented a CFD simulation of the blood flow in the retinal arteries of a planar 2D vessel network model. The vessel network with 52 outlets and visible branches with diameters in the range of 14–120 μm were obtained from high-resolution fundus photograph with the Frangi filter method. This is the most comprehensive human retinal artery network reported in the literature, where a CFD simulation was performed. Based on this initial model, further CFD studies can be realised to better understand the hemodynamics in the retinal microcirculation.

References

1. ANSYS, Inc, FLUENT 14.5 User Guide (2014)
2. D.L. Brown, R. Cortez, M.L. Minion, Accurate projection methods for the incompressible Navier–Stokes equations. J. Comput. Phys. **168**(2), 464–499 (2001)
3. A. Budai, J. Odstricilik, R. Kollar, J. Jan, T. Kubena, G. Michelson, A public database for the evaluation of fundus image segmentation algorithms, Poster at Fort Lauderdale Convention Center, The Association of Research in Vision and Ophthalmology (ARVO) Annual Meeting in Fort Lauderdale (2011)
4. A. Budai, R. Bock, A. Maier, J. Hornegger, G. Michelson, Robust vessel segmentation in fundus images. Int. J. Biomed. Imaging **2013**, 1–11 (2013)
5. Y.I. Cho, K.R. Kensey, Effects of the non-Newtonian viscosity of blood on flows in a diseased arterial vessel. Part 1: Steady flows. Biorheology **28**(3–4), 241–262 (1991)
6. G. Clough, J.L. Cracowski, Spotlight issue: microcirculation-from a clinical perspective. Microcirculation **19**(1), 1–4 (2012)
7. M. Day, The no-slip condition of fluid dynamics. Erkenntnis **33**(3), 285–296 (1990)
8. T. Gracner, Ocular blood flow velocity determined by color Doppler imaging in diabetic retinopathy. Ophthalmologica **218**(4), 237–242 (2004)
9. G. Guidoboni, A. Harris, S. Cassani, J. Arciero, B. Siesky, A. Amireskandari, L. Tobe, P. Egan, I. Januleviciene, J. Park, Intraocular pressure, blood pressure, and retinal blood flow autoregulation: a mathematical model to clarify their relationship and clinical relevance. Invest. Ophthalmol. Vis. Sci. **55**(7), 4105–4118 (2014)
10. L.D. Hubbard, R.J. Brothers, W.N. King, L.X. Clegg, R. Klein, L.S. Cooper, A.R. Sharrett, M.D. Davis, J. Cai, Methods for evaluation of retinal microvascular abnormalities associated with hypertension/sclerosis in the atherosclerosis risk in communities study. Ophthalmology **106**(12), 2269–2280 (1999)
11. H. Kolb, E. Fernandez, R. Nelson, *Facts and Figures Concerning the Human Retina in Webvision: The Organization of the Retina and Visual System* (University of Utah, Salt Lake City, 1995)
12. C. Kondermann, D. Kondermann, M. Yan, Blood vessel classification into arteries and veins in retinal images, in *Conference Proceedings* (2007), pp. 651247–651249

13. G. Liew, M.L. Baker, T.Y. Wong, P.J. Hand, J.J. Wang, P. Mitchell, D.A. De Silva, M.C. Wong, E. Rochtchina, R.I. Lindley, J.M. Wardlaw, G.J. Hankey, Multi-Centre Retinal Stroke Study Group, Differing associations of white matter lesions and lacunar infarction with retinal microvascular signs. Int. J. Stroke **9**(7), 921–5 (2014)
14. D. Liu, N.B. Wood, N. Witt, A.D. Hughes, S.A. Thom, X.Y. Xu, Computational analysis of oxygen transport in the retinal arterial network. Curr. Eye Res. **34**(11), 945–956 (2009)
15. J. Malek, T. Azar, B. Nasralli, M. Tekari, H. Kamoun, R. Tourki, Computational analysis of blood flow in the retinal arteries and veins using fundus image. Comput. Math. Appl. **69**, 101–116 (2015)
16. M. Martinez-Perez, Computer analysis of the geometry of the retinal vasculature. Ph.D. thesis, Imperial College London, 2000
17. A. Mendivil, V. Cuartero, Color Doppler echography study of ocular blood flow velocity in patients with proliferative diabetic retinopathy after performance of retinal panphotocoagulation: 2 years' follow-up. Rev. Med. Univ. Navarra **42**(3), 134–44 (1998)
18. C.D. Murray, The physiological principle of minimum work: I. The vascular system and the cost of blood volume. Proc. Natl. Acad. Sci. USA **12**(3), 207–14 (1926)
19. M.S. Olufsen, Structured tree outflow condition for blood flow in larger systemic arteries. Am. J. Physiol. Heart Circ. Physiol. **276**(1 45-1), 257–268 (1999)
20. M.S. Olufsen, C.S. Peskin, W.Y. Kim, E.M. Pedersen, A. Nadim, J. Larsen, Numerical simulation and experimental validation of blood flow in arteries with structured-tree outflow conditions. Ann. Biomed. Eng. **28**(11), 1281–1299 (2000)
21. H.D. Papenfuss, J.F. Gross, Microhemodynamics of capillary networks. Biorheology **18**(3–6), 673–692 (1981)
22. A.R. Pries, T.W. Secomb, P. Gaehtgens, J.F. Gross, Blood flow in microvascular networks. Experiments and simulation. Circ. Res. **67**(4), 826–834 (1990)
23. A.R. Pries, T.W. Secomb, P. Gaehtgens, Biophysical aspects of blood flow in the microvasculature. Cardiovasc. Res. **32**(4), 654–667 (1996)
24. M. Quigley, S. Cohen, A new pressure attenuation index to evaluate retinal circulation. A link to protective factors in diabetic retinopathy. Arch. Ophthalmol. **117**(1), 84–89 (1999)
25. C.E. Riva, J.E. Grunwald, S.H. Sinclair, B.L. Petrig, Blood velocity and volumetric flow rate in human retinal vessels. Invest. Ophthalmol. Vis. Sci. **26**(8), 1124–1132 (1985)
26. M.E. Safar, P. Lacolley, Disturbance of macro- and microcirculation: relations with pulse pressure and cardiac organ damage. Am. J. Physiol. Heart Circ. Physiol. **293**(1), H1–H7 (2007)
27. A.K. Schuster, J.E. Fischer, U. Vossmerbaeumer, A retinal snap shot may indicate individual risk for cardiovascular disease - the MIPH eye&health study. Int. J. Cardiol. **180**, 30–33 (2015)
28. B.N. Steele, M.S. Olufsen, C.A. Taylor, Fractal network model for simulating abdominal and lower extremity blood flow during resting and exercise conditions. Comput. Methods Biomech. Biomed. Eng. **10**(1), 39–51 (2007)
29. T. Takahashi, T. Nagaoka, H. Yanagida, T. Saitoh, A. Kamiya, T. Hein, L. Kuo, A. Yoshida, A mathematical model for the distribution of hemodynamic parameters in the human retinal microvascular network. J. Biorheol. **23**(2), 77–86 (2009)
30. B.N. Tillmann, *Atlas der Anatomie* (Springer, Heidelberg, 2010)
31. Y. Wang, B.A. Bower, J.A. Izatt, O. Tan, D. Huang, In vivo total retinal blood flow measurement by Fourier domain Doppler optical coherence tomography. J. Biomed. Opt. **12**(4), 041215–041218 (2007)
32. W. Wells, A. Colchester, S. Delp, A. Frangi, W. Niessen, K. Vincken, M. Viergever, *Multiscale Vessel Enhancement Filtering*. Lecture Notes in Computer Science, vol. 1496 (Springer, Berlin/Heidelberg, 1998), pp. 130–137
33. M. Zamir, Shear forces and blood vessel radii in the cardiovascular system. J. Gen. Physiol. **69**(4), 449–461 (1977)

Chapter 14
Integration of an Electrophysiologically Driven Heart Model into Three-Dimensional Haemodynamics Simulation Using the CRIMSON Control Systems Framework

Christopher J. Arthurs and C. Alberto Figueroa

14.1 Introduction

Lumped parameter network (LPN) models have been used extensively to simulate behaviour within the cardiovascular system, either exclusively [3, 29] or coupled with one-dimensional [1, 12, 13, 18] or three-dimensional [17, 24, 25, 30] vascular domains. Vascular regions which have been investigated using LPN models include the coronary arteries [7, 8], the heart [5, 9, 13, 14], the brain [1] and full closed-loop simulations [17]. With a few exceptions [1, 6], these LPNs have used static parameters. An awareness of the deficiencies of purely static or steady-state simulation in computational haemodynamics means that there is interest in models which can adjust their own parameters in a physiologically inspired manner. Recent examples include using an autonomic nervous system reflex to control cardiac parameters [9], or ensuring that oxygen delivery to the myocardium closely matches cardiac metabolic demand [2]. Controlled models are important not only because they reproduce key phenomena such as the change in heart rate when standing up [9], or the changes in coronary flow that occur during exercise [2], but also because

C.J. Arthurs (✉)
Division of Imaging Sciences and Biomedical Engineering, King's College London, King's Health Partners, St. Thomas' Hospital, London SE1 7EH, UK
e-mail: christopher.arthurs@kcl.ac.uk

C.A. Figueroa
Department of Surgery, University of Michigan, Ann Arbor, MI 48109, USA

Department of Biomedical Engineering, University of Michigan, Ann Arbor, MI 48109, USA

Division of Imaging Sciences and Biomedical Engineering, King's College London, King's Health Partners, St. Thomas' Hospital, London SE1 7EH, UK
e-mail: figueroc@med.umich.edu

G. Joldes et al. (eds.), *Computational Biomechanics for Medicine*,
DOI 10.1007/978-3-319-28329-6_14

the study of the highly integrated networks that cardiovascular control systems form is challenging in vivo or in vitro, from both the technical and the conceptual perspective.

Despite the accepted need for control systems models, progress is hindered by the time-consuming nature of implementing and testing control systems within existing powerful simulation packages. In this work, we present the latest developments made to our cardiovascular geometry creation and incompressible Navier–Stokes haemodynamics simulation software, CRIMSON (CardiovasculaR Integrated Modelling and SimulatiON) [28], which we assert can accelerate progress by making the design of controlled physiological models faster, easier and more accessible, even to users without a strong background in software development. In order to demonstrate their flexibility, we use CRIMSON's boundary condition and control system design tools, the CRIMSON Boundary Condition Toolbox (BCT) and the CRIMSON Control Systems Framework (CSF), to create an electrophysiologically driven heart model, and use it as an inflow boundary condition as part of a multidomain, multiphysics Navier–Stokes haemodynamics simulation in an example vascular geometry. The model makes use of an existing biophysical model of the cardiac myocyte and its active tension generation, which we obtain from the mathematical cell model repository cellML [10, 27]. The benefit of using an electrophysiologically driven heart model is that it allows us to leverage decades of modelling work on the behaviour of the cardiac myocyte, with different desirable properties available depending on the particular choice of myocyte model. The primary purpose of this article is to demonstrate rapid model design and integration, so our heart model follows previous work [19].

Previous non-electrophysiological LPN heart models generally employ a time-varying elastance method [3, 16, 21], and include those that model flow-rate dependent pressure losses in the left ventricle [9]. These models successfully reproduce aortic pressure and flow waveforms. Electrophysiological LPN models have been shown to allow the effect of subcellular processes upon the haemodynamics to be investigated, for example, aortic pressure can be seen to depend on L-type calcium channel conductance and upon on pacing frequency [19]. Similarly, appropriate electrophysiological heart models are capable of reproducing the Frank–Starling mechanism, the effects of dyssynchronous contraction and choice of pacing location [26]. In general, the use of the model means that the aortic valve inflow in the fluid domain is dependent upon subcellular processes and parameters, including transmembrane voltage difference, ion channel state and intracellular calcium concentration.

14.2 Methods

14.2.1 Overview of CRIMSON

We perform our simulations using CRIMSON, which provides a complete software pipeline for creating Navier–Stokes haemodynamics simulations from medical imaging stacks, with an emphasis on power and usability. It consists of two main components: the intuitive image analysis and segmentation interface, and the powerful flowsolver simulation package. The flowsolver is highly scalable, having been used previously to simulate pulsatile flow on 16,384 cores of an IBM Blue Gene/Q supercomputer. In the present work we discuss only two aspects of the pipeline: the boundary condition control tool: CRIMSON CSF, and the closely related arbitrary LPN condition design and specification tool: CRIMSON BCT. We used CRIMSON to create a simple vessel geometry for our investigations, which can be seen in Fig. 14.2.

14.2.2 Graphical Design of Arbitrary Lumped Parameter Boundary Conditions

Lumped parameter components are assembled into a network using a drag-n-drop interface. The available components include resistances, compliances, valves, inductances and volume-tracking compliance chambers, and they can be arbitrarily arranged, and attached at a point to a boundary of the 3D simulation domain, as shown in Fig. 14.1. One circuit is created for each boundary, connected, for example, as shown in Fig. 14.2, and if desired, a circuit to represent the venous system can be created and attached to some or all of the boundary circuits, in order to create a full closed-loop network. We used the BCT to create the heart model (Fig. 14.3) and two downstream Windkessel models.

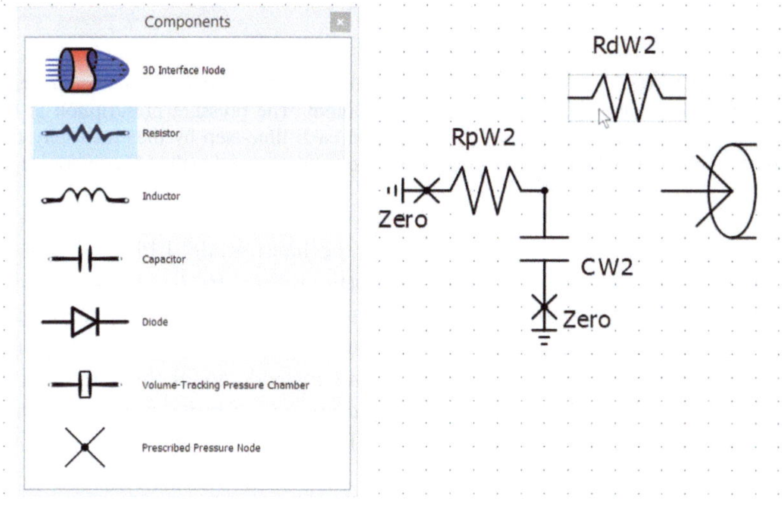

Fig. 14.1 The creation of a three-element Windkessel model using the drag-n-drop CRIMSON Boundary Condition Toolbox

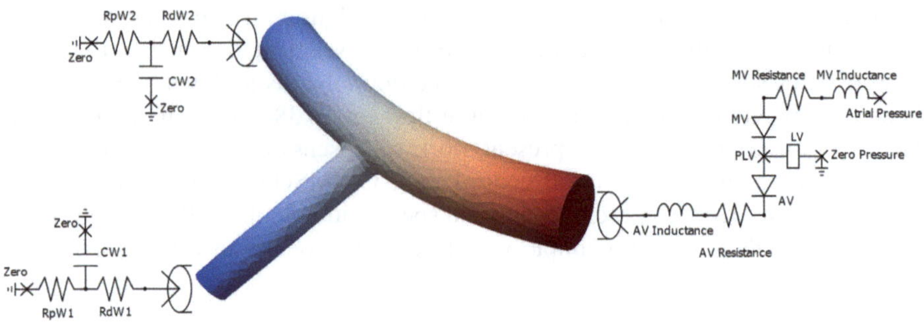

Fig. 14.2 A complete set of boundary conditions, designed in the arbitrary boundary condition toolbox and attached to a 3D domain. A heart model is shown on the *right*, and the three-element Windkessel models on the *left* represent downstream vascular beds

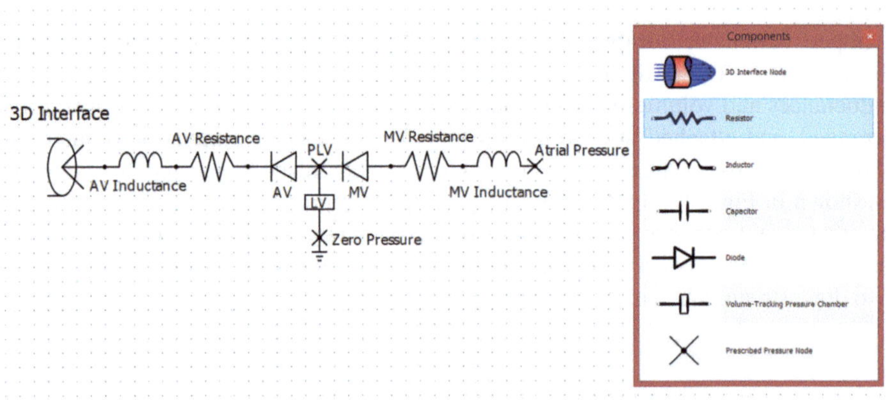

Fig. 14.3 The CRIMSON Boundary Condition Toolbox, used here to design a heart model. Nodes with prescribed pressure are tagged with the X symbol. The pressure prescription at the left-ventricular pressure node (PLV) here will be set on each time-step by the electrophysiological cell model, within the Python control script, using the volume stored in the LV component as input

We use a standard component layout for the heart model LPN; similar designs have been used previously to simulate aortic inflow [6, 9]. However, because we want to control pressure generation using an electrophysiological model, we abandon the usual feature that the pressure within the ventricle is computed using a time-varying elastance approach, and instead model the left ventricle by a component which simply keeps track of the volume of blood that it contains. The construction of the model in CRIMSON BCT is shown in Fig. 14.3.

14.2.3 Powerful, Rapid and Accessible Control Systems Design

CRIMSON CSF has access to, and is able to adjust, any of the component parameters or nodal pressures within any of the CRIMSON BCT boundary condition models, and it has access to all of the pressures, volumes and flows within each boundary condition. This provides sufficient functionality for modelling many physiological control mechanisms. Control systems themselves are described using Python, a popular high-level language which is suitable for both beginners and advanced users. While the CRIMSON flowsolver itself is written in Fortran and C++, both of which require considerable expertise to work with, Python is similar to MATLAB in terms of being much easier to learn and to use. The Python interface with the CRIMSON flowsolver is simple, and works as follows. To design a control system for a particular node or component, we annotate it with the name of the Python controller script within the arbitrary boundary condition toolbox. We then take the CRIMSON Python script template which contains all of the necessary boilerplate code, including the automatically passed-in data on the state of the system, and the return value (the new value of the controlled parameter that we wish to set), and we write the code for the custom control system we wish to design into the template. A simple example control script for controlling a resistor is shown in Fig. 14.4.

```
from math import pi, cos

# Each controller is a class
class sinusoidalResistanceController:

    # This function is called once, at the start of the simulation.
    # We can use it to set up any constant values we might need
    def __init__(self):
        self.m_periodicTime = 0.0;
        self.m_heartPeriod = 0.86;


    # updateControl is the method that the CRIMSON flowsolver looks for. It calls it
    # on each timestep, and passes all the pressures, flows and stored volumes in
    # the attached boundary condition. The flowsolver it knows what to do with the
    # return value; in this case, it will use it to set a resistance, causing
    # it to vary in time.
    def updateControl(self, currentResistance, timestep, pressuresInBoundaryCondition, \
                      flowsInBoundaryCondition, volumesInBoundaryCondition):
        # Call the function defined below to update the current time
        self.updatePeriodicTime(timestep)
        # Cause the resistance of the target resistor to oscillate in time, and to depend
        # on the pressure at the node with index 1 within the boundary condition
        resistance = cos(self.m_periodicTime * 2.0 * pi) + 5.0 + pressuresInBoundaryCondition[1]
        # Return the resistance
        return resistance


    # We can write additional functions to call from updateControl
    def updatePeriodicTime(self, timestep):
        self.m_periodicTime = self.m_periodicTime + timestep
        # Keep m_periodicTime in the range [0,m_heartPeriod):
        if self.m_periodicTime >= self.m_heartPeriod:
            self.m_periodicTime = self.m_periodicTime - self.m_heartPeriod

```

Fig. 14.4 A simple example Python script that could be used to control a resistance in one of the boundary conditions, dependent here on time and on a pressure within the boundary condition

14.2.4 The Cardiac Myocyte Model

The use of the electrophysiological model follows a previous approach used for zero-dimensional simulation due to Shim et al. [19], in which the authors took an existing electrophysiological model of a cardiac myocyte, the ten Tusscher 2004 model [23] which, upon the application of an electrical stimulus, generates an action potential caused by the flow of ions across the cell membrane. The model simulates the concomitant calcium release within the cell, which is the internal signal which causes the cell to generate active tension. This model was modified to be suitable for connection to an intracellular cross-bridge dynamics model [15], which generates the active tension in response to the calcium release. From this, Bo Shim et al. created a pressure generation model by assuming the ventricle to be a thin-walled hemispherical shell, and applying Laplace's law to convert a known volume and wall tension into ventricular pressure. Our approach uses the electrophysiological model of Shirokov et al. [20] coupled with the Negroni and Lascano model for active tension generation, as this combination was available in the cellML [10, 27] repository, as the work of Matsuoka et al. [11]. We modified the model to include the thin-shell-based ventricular pressure generation approach of Bo Shim et al.

14.2.5 Inserting the Cardiac Cell Model into the CRIMSON Flowsolver Using the Control Systems Framework

We downloaded the Matsuoka model from the cellML model repository as Python code [4], and inserted it into our CRIMSON–Python interface template script, modifying it so that it would advance a single time-step each time it was called to update the control. We did not adjust the parameters from the cellML exposure of the model [4]. We introduced the shell-based pressure computation, using the left-ventricular volume data automatically passed to the controller by the flowsolver, and including the change of half-sarcomere length as the myocytes are stretched by the volume within the ventricle. We further modified the model so that during diastole, the filling is controlled by a constant diastolic elastance. The computed pressure is returned to the flowsolver at the end of each update, and is used to set the left-ventricular pressure within the heart model. When this prescribed pressure exceeds the aortic pressure, the aortic valve opens and blood flows into the aorta, and the volume in the left ventricle seen by the control system is reduced. The converse is true during diastolic filling.

We relate active tension to left-ventricular pressure using the left-ventricular volume, a spherical approximation of the ventricle, and Laplace's law, and the parameters of the Windkessel models at the two other boundaries of the domain, seen in Fig. 14.2, were tuned to adjust the aortic pressure waveform and ventricular ejection fraction.

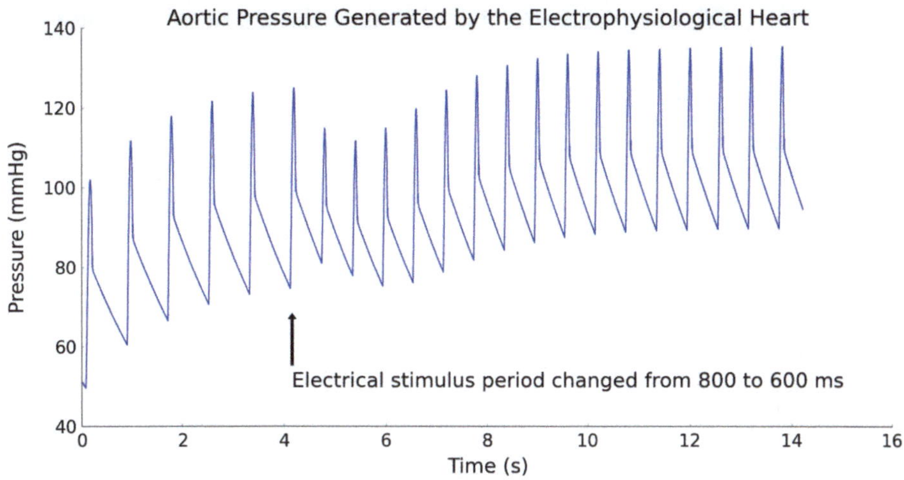

Fig. 14.5 Aortic pressure generated by the heart model. Note that changes in heart rate are handled automatically and naturally by the cardiac cell model

14.3 Results

We were able to achieve our primary objective of creating a complex boundary condition, an electrophysiological heart model, by making use of the CRIMSON Boundary Condition Toolbox and Control Systems Framework. This demonstrates the power of the tools, which enabled us to create the heart model from initial design to full functionality in the space of 2 days.

Figures 14.5 and 14.6 show that the heart model successfully reproduces an aortic pressure pattern and left-ventricular pressure–volume loop. Each beat is the result of an electrical stimulus applied to the myocyte within the control script, and so we can change the heart rate by changing the frequency of the electrical stimulus; we do this 4 s into the simulation shown in Fig. 14.5.

14.4 Discussion

We successfully used the model to generate inflow pressure and flow in a Navier–Stokes simulation domain, and to generate pressure–volume loops for the left ventricle. The use of the cell model allows us to initiate each pressure pulse by simulating the application of an electrical stimulus. The development of the model was rapid, due to the novel tools which we have created.

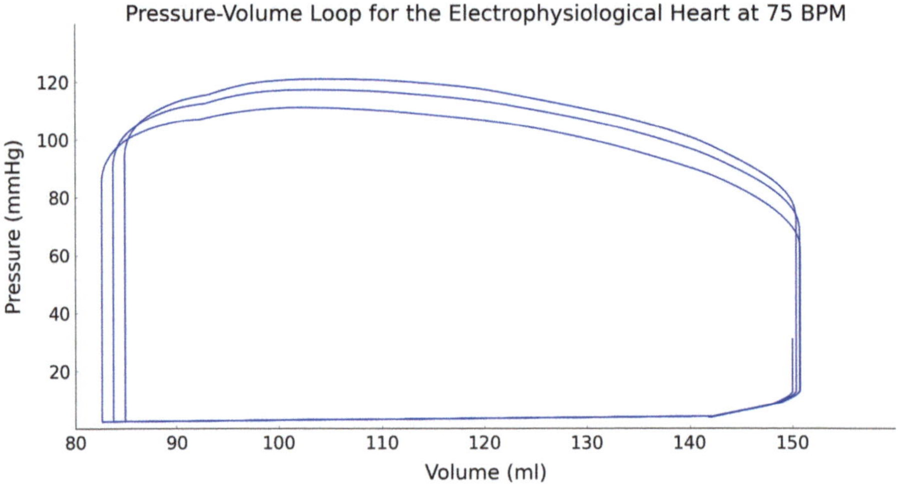

Fig. 14.6 The pressure–volume loop produced within the left ventricle by the electrophysiological heart model, implemented using the control systems framework. Several beats are displayed

14.4.1 The Cardiac Cell Model

Using a cardiac electrophysiology myocyte model means that the effects of changing the electrical pacing cycle length on the cell's internal state variables are naturally propagated to the generation of ventricular pressure. Cell models have differing levels of realism in their ability to reproduce physiologically observed phenomena. The model of Shirokov et al. [20], as modified by Matsuoka et al. [11], is only one such possibility. One reason to investigate other modes is that the duration of systole is too short. This is a limitation of the cell model used; it is likely caused by the Matsuoka model using data from guinea pig myocytes. This is something that we could improve upon by replacing the electrophysiological component of the Matsuoka model with one for a human myocyte [22, 23].

14.4.2 Scope of Arbitrary Cardiovascular Control Mechanism Design

CRIMSON CSF aims to provide a complete set of tools for controlling the parameters within boundary condition models. For example, its access to the parameters which determine physiologically important factors such as tissue perfusion and oxygen delivery means that control systems which monitor and adjust to varying tissue perfusion requirements can be created. Additionally, control systems which do not rely on any such monitoring, such as the cardiovascular response to psychological stress, could be simulated by creating a control system which does not use any of these variables as input. We believe that the facility to in this manner adjust any of the nodal pressures and any of the component parameters within the boundary conditions should allow most physiological control systems to be modelled.

14.4.3 Rapid Prototyping

One of the features which we found to be the most useful during this work was the facility for rapid boundary condition design, testing and approximate parameterization provided by the CRIMSON flowsolver's pure zero-dimensional prototyping mode. Enabled using a single input flag, this mode automatically replaces the 3D simulation domain with an additional, simplified zero-dimensional domain (Fig. 14.7), allowing many hundreds of cardiac cycles to be simulated in a short period of time on a laptop, as opposed to achieving a few beats per hour on powerful computing hardware. This is particularly useful for approximately parametrizing a control system in order to study some state transition, as we generally require the system to reach an equilibrium state before testing a control perturbation, and then we require a further extended period of simulation to observe

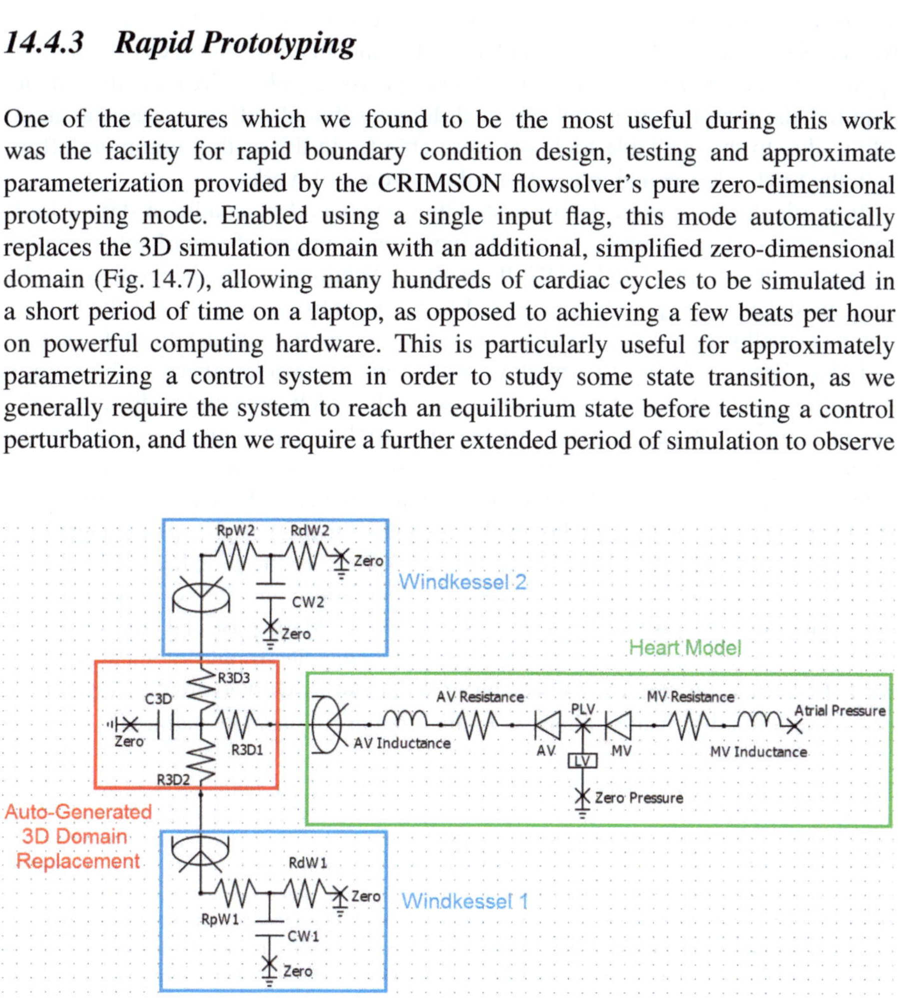

Fig. 14.7 A schematic of circuit in pure zero-dimensional mode. In this mode, CRIMSON flowsolver automatically generates a replacement for the 3D domain (*red*) with the same topology (compare Fig. 14.2), and connects it to the boundary conditions, as prescribed for the 3D interface. This allows very rapid prototyping simulations to be run

the transitional behaviour. We note that because this mode neglects all 3D effects, the resulting parameterization should only be seen as an approximate value, which must be fine-tuned in full 3D simulation mode.

14.5 Conclusions

We performed multiphysics simulation of the cardiovascular system by using an electrophysiological heart model to generate flow within a three-dimensional Navier–Stokes haemodynamics simulation. The model allows an electrical stimulus applied to the myocyte to trigger a blood pressure pulse. Creating this model required the merging of models from different subfields of cardiovascular modelling; due to the available tools we were able to do this with a minimum effort, with the model design and integration taking 2 days of work.

In particular, this work demonstrates that our boundary condition design tools and control systems framework enable rapid development of remarkably complex enhancements of the CRIMSON flowsolver. While pressure generation in the heart model is not typically considered to be a control system, using the control framework allowed us to show that it is useful for more than just control systems, and also, because fusing two models in this manner would otherwise be a time-consuming task, it demonstrated the ease with which potentially difficult tasks can be achieved. A key purpose of these new tools is that it gives researchers the space to explore, so we do not expect to predict all possible uses, but to list a few, potential applications include simulating haemorrhage, both by creating the bleed in the first place and by simulating the response of the peripheral resistance and venous compliance, modelling the exercise response in the peripheral vasculature and in the heart, as coordinated by the neural central command, or implementing autoregulation systems within individual tissue beds. Many potential control system models will have a lower level of complexity than the electrophysiological heart, so we believe that our framework will be of great use to workers as they design the next generation of transitional physiological models in haemodynamics.

Acknowledgements We gratefully acknowledge support from the European Research Council under the European Union's Seventh Framework Programme (FP/2007-2013)/ERC Grant Agreement n. 307532, and the United Kingdom Department of Health via the National Institute for Health Research (NIHR) comprehensive Biomedical Research Centre award to Guy's and St Thomas' NHS Foundation Trust in partnership with King's College London and King's College Hospital NHS Foundation Trust. The authors wish to thank Rostislav Khlebnikov and Kevin Lau for their assistance and advice during this work. We acknowledge the developers of QSapecNG (http://qsapecng.sourceforge.net/), the GPLv3-licensed circuit simulation package which we have modified to create the CRIMSON arbitrary boundary condition toolbox, and the Free Software Foundation (https://www.fsf.org/) for supporting innovation via GNU (https://www.gnu.org/home.en.html) and the GPL Free Software Licenses (https://www.gnu.org/licenses/gpl-3.0.en.html).

References

1. J. Alastruey, S.M. Moore, K.H. Parker, T. David, J. Peirò, S.J. Sherwin, Reduced modelling of blood flow in the cerebral circulation: coupling 1-D, 0-D and cerebral auto-regulation models. Int. J. Numer. Methods Fluids **56**, 1061–1067 (2008)
2. C.J. Arthurs, K.D. Lau, K.N. Asrress, S. Redwood, C.A. Figueroa, A mathematical model of coronary blood flow control: patient-specific three-dimensional haemodynamics during exercise. Am. J. Physiol. Heart Circ. Physiol. (2016, in press)
3. O. Barnea, Mathematical analysis of coronary autoregulation and vascular reserve in closed-loop circulation. Comput. Biomed. Res. **27**, 263–275 (1994)
4. A. Garny, CellML Model Exposure: role of individual ionic current systems in the SA node hypothesized by a model study, Matsuoka, Sarai, Kuratomi, Ono, Noma, 2003 (2009), https://models.cellml.org/exposure/398d5dc7db9f2b9809abc29f440bd456/. Accessed 20 May 2015 at 18:42 BST
5. H.J. Kim, I.E. Vignon-Clementel, C.A. Figueroa, J.F. LaDisa, K.E. Jansen, J.A. Feinstein, C.A. Taylor, On coupling a lumped parameter heart model and a three-dimensional finite element aorta model. Ann. Biomed. Eng. **37**, 2153–2169 (2009)
6. H.J. Kim, K.E. Jansen, C.A. Taylor, Incorporating autoregulatory mechanisms of the cardiovascular system in three-dimensional finite element models of arterial blood flow. Ann. Biomed. Eng. **38**, 2314–2330 (2010)
7. H.J. Kim, I.E. Vignon-Clementel, J.S. Coogan, C.A. Figueroa, K.E. Jansen, C.A. Taylor, Patient-specific modeling of blood flow and pressure in human coronary arteries. Ann. Biomed. Eng. **38**, 3195–3209 (2010)
8. H.J. Kim, I.E. Vignon-Clementel, C.A. Figueroa, K.E. Jansen, C.A. Taylor, Developing computational methods for three-dimensional finite element simulations of coronary blood flow. Finite Elem. Anal. Des. **46**, 514–525 (2010)
9. K.D. Lau, C.A. Figueroa, Simulation of short-term pressure regulation during the tilt test in a coupled 3D–0D closed-loop model of the circulation. Biomech. Model. Mechanobiol. **14**, 915–929 (2015)
10. C.M. Lloyd, J.L. Lawson, P.J. Hunter, P.F. Nielsen, The CellML model repository. Bioinformatics **24**, 2122–2123 (2008)
11. S. Matsuoka, N. Sarai, S. Kuratomi, K. Ono, A. Noma, Role of individual ionic current systems in ventricular cells hypothesized by a model study. Jpn. J. Physiol. **53**, 105–123 (2003)
12. K.S. Matthys, J. Alastruey, J. Peiró, A.W. Khir, P. Segers, P.R. Verdonck, K.H. Parker, S.J. Sherwin, Pulse wave propagation in a model of human arterial network: assessment of 1-D numerical simulations against in vitro measurements. J. Biomech. **40**, 3476–3486 (2007)
13. J. Mynard, P. Nithiarasu, A 1D arterial model incorporating ventricular pressure, aortic valve and regional coronary flow using the locally conservative Galerkin (LCG) method. Commun. Numer. Methods Eng. **24**, 367–417 (2008)
14. J. Mynard, M.R. Davidson, D.J. Penny, J.J. Smolich, A simple, versatile valve model for use in lumped parameter and one-dimensional cardiovascular models. Int. J. Numer. Methods Biomed. Eng. **28**, 626–641 (2012)
15. J. Negroni, E.C. Lascano, A cardiac muscle model relating sarcomere dynamics to calcium kinetics. J. Mol. Cell. Cardiol. **28**, 915–929 (1996)
16. S.R. Pope, L.M. Ellweian, C.L. Zapata, V. Novak, C.T. Kelley, M.S. Olufsen, Estimation and identification of parameters in a lumped cerebrovascular model. Math. Biosci. Eng. **6**, 93–115 (2009)
17. A. Quarteroni, S. Ragni, A. Veneziani, Coupling between lumped and distributed models for blood flow problems. Comput. Vis. Sci. **4**, 111–124 (2001)
18. P. Raymond, F. Merenda, F. Perren, D. Rüfenacht, N. Stergiopulos, Validation of a one-dimensional model of the systemic arterial tree. Am. J. Physiol. Heart Circ. Physiol. **297**, H208–H222 (2009)

19. E.B. Shim, C.H. Leem, Y. Abe, A. Noma, A new multi-scale simulation model of the circulation: from cells to system. Philos. Trans. R. Soc. A **364**, 1483–1500 (2006)
20. R. Shirokov, R. Levis, N. Shirokova, E. Riós, Ca2+-dependent inactivation of l-type Ca2+ channels does not affect their voltage sensor. J. Gen. Physiol. **102**, 1005–1030 (1993)
21. H. Suga, K. Sagawa, Instantaneous pressure-volume relationships and their ratio in excised, supported canine left ventricle. Circ. Res. **35**, 117–126 (1974)
22. K.H.W.J. ten Tusscher, A.V. Panfilov, Alternans and spiral breakup in a human ventricular tissue model. Am. J. Physiol. Heart Circ. Physiol. **291**, H1088–H1100 (2006)
23. K.H.W.J. ten Tusscher, D. Noble, P.J. Noble, A.V. Panfilov, A model for human ventricular tissue. Am. J. Physiol. Heart Circ. Physiol. **286**, H1573–H1589 (2004)
24. I.E. Vignon-Clementel, A coupled multidomain method for computational modeling of blood flow. Ph.D. thesis, Stanford University, 2006
25. I.E. Vignon-Clementel, C.A. Figueroa, K.E. Jansen, C.A. Taylor, Outflow boundary conditions for the three-dimensional finite element model of blood flow and pressure in arteries. Comput. Methods Appl. Mech. Eng. **195**, 3776–3796 (2006)
26. J. Walmsley, T. Arts, N. Derval, P. Bordachar, H. Cochet, S. Ploux, F.W. Prinzen, T. Delhaas, J. Lumens, Fast simulation of mechanical heterogeneity in the electrically asynchronous heart using the multipatch module. PLoS Comput. Biol. (2015). doi:10.1371/journal.pcbi.1004284
27. Website, CellML Model Repository, Auckland Bioengineering Institute, University of Auckland (2015), https://www.cellml.org/. Accessed 22 July 2015
28. Website, CRIMSON (CardiovasculaR Integrated Modelling and SimulatiON) (2015), http://www.crimson.software. Accessed 22 July 2015
29. N. Westerhof, F. Bosman, C.J. de Vries, A. Noordergraaf, Analog studies of the human systemic arterial tree. J. Biomech. **2**, 121–143 (1969)
30. N. Xiao, J. Humphrey, C. Figueroa, Multi-scale computational model of three-dimensional hemodynamics with a deformable full-body arterial network. J. Comput. Phys. **224**, 22–40 (2013)

Chapter 15
Simulating Patient Specific Multiple Time-Point MRIs from a Biophysical Model of Brain Deformation in Alzheimer's Disease

Bishesh Khanal, Marco Lorenzi, Nicholas Ayache, and Xavier Pennec

15.1 Introduction

Alzheimer's Disease (AD) is one of the most common types of dementia. It is a neurodegenerative disease that progresses gradually over several years with the accumulation of neurofibrillary tangles (NFTs) and amyloid-β (A-β) plaques [2]. These microscopic neurobiological changes are followed by the progressive neuronal damage that leads to the atrophy of the brain tissue. The atrophy or the volume changes of brain tissue are a macroscopic change that structural Magnetic Resonance Imaging (MRI) can estimate in different brain regions. Many different methods have been proposed to estimate atrophy in some particular regions of brain that are known to be affected in AD [9].

In addition to estimating specific brain structures with atrophy, longitudinal imaging data could also potentially be used to study the temporal inter-relationship of atrophy in different structures. For instance, in [6], authors estimate per-individual rates of atrophy in 34 cortical regions and in hippocampus. Then they study the groupings of these structures based on the correlation of the atrophy rates. In [8], authors define AD progression as a series of discrete events. Atrophy in different parts of the brain is taken as different events along with clinical events. Without any prior to their ordering, the model finds the most probable order for these events from the data itself. They use Bayesian statistical algorithms for fitting in the

B. Khanal (✉) • N. Ayache • X. Pennec
Inria Sophia Antipolis Méditerranée, Asclepios Research Project, 2004 Route des Lucioles - BP 93, 06902 Sophia Antipolis, France
e-mail: bishesh.khanal@inria.fr

M. Lorenzi
University College London, Translational Imaging Group, London, UK

INRIA Sophia Antipolis Méditerranée, Asclepios Research Project, 2004 Route des Lucioles - BP 93, 06902 Sophia Antipolis, France

G. Joldes et al. (eds.), *Computational Biomechanics for Medicine*,
DOI 10.1007/978-3-319-28329-6_15

event-based disease progression model. The objective of these kinds of studies is to understand how different regions of brain interact during the neurodegeneration and find its trajectory. Such studies can benefit with large number of longitudinal images of AD patients. In this context, a model that can simulate many time-point images from a few available longitudinal images can be a valuable tool.

Atrophy simulators [4, 11, 14, 18] have been proposed in the literature and used mostly for the validation of registration or segmentation methods [5, 16], or to estimate uncertainty in the measured atrophy [17]. The simulators in [11, 14, 16] use a Jacobian based method where the desired level of atrophy is set at each voxel, and the deformation that best approximates the desired level of atrophy is found. Regularization is used in the optimization to enforce certain desired conditions such as topology preservation. The advantage of these methods is the ability to define atrophy maps at the voxel level. However regularization parameters used to enforce topology preservation are generally difficult to relate to a plausible biophysical process of AD and can create difficulties in simulating opening of certain structures such as sulci. It is not trivial to consider different tissue behaviors in such approaches. In [4, 18], authors propose a model of brain deformation based on thermoelasticity. Volume changes are defined in particular structures/tissues of a meshed brain by assigning different thermal coefficients. Simulation of the images is done by first solving the thermoelastic model of tissue deformation with Finite Element Method (FEM), and then by interpolating the obtained displacement field from the mesh to the image. FEM involves moving back and forth from voxels to meshes which creates numerical difficulties and inaccuracies in the model personalization.

In [12] we proposed a new biophysical model of brain deformation due to atrophy in AD that combines the advantages of the models mentioned in the previous paragraph. The mechanisms of neuronal deaths and its evolution are not well known for AD and are likely to be primarily guided by complex physiological processes. However we believe that the biomechanics of brain tissue might play an important role in determining the consequence of the neuronal deaths on brain shape changes. Our biophysical model presented in [12] builds upon the assumptions that we relate to the biophysical process of tissue shape changes as the consequence of local volume loss. This model can be used to simulate time-series MRIs starting from a real input baseline MRI.

In this work we use our biophysical model developed in [12] to present a framework that allows to interpolate or extrapolate patient specific unseen time-point images from at least two available time-point images of the subject and to assess how closely these simulated trajectories follow real patient trajectories. We also improve the implementation of the boundary condition of the model by imposing zero deformation in the skull and all the regions outside of the skull. In [12] the zero deformation was imposed at the image boundaries and not at the brain–skull boundary.

The following section briefly explains the assumptions and implementation of the biophysical model we presented in [12], and in Sect. 15.3 we present how we interpolate new images between two acquisition time points.

15.2 Biophysical Model of Brain Deformation Due to Atrophy

The atrophy rate $\tilde{a}(\boldsymbol{x}, t)$ at any position $\boldsymbol{x}$ at time t for a representative elementary volume of $V(\boldsymbol{x}, t)$ is defined as the negative rate of change of volume per unit volume:

$$\boxed{\tilde{a} = \frac{-1}{V}\frac{\partial V}{\partial t}.}$$

For any time Δt that results in sufficiently small displacement, the amount of atrophy is $a = \tilde{a}\Delta t$. Any deformation field that has atrophy a should satisfy the following equation:

$$\nabla \cdot \boldsymbol{u} = -a, \tag{15.1}$$

where u is the displacement of material particles during the Δt.

We do not explicitly model the neuronal loss and tissue remodeling at the microscopic level which requires biochemical and cellular physiological knowledge in detail. We abstract the phenomenon that evolves over several months or years in the brain. In Creutzfeldt–Jakob disease, no gross brain shape changes are reported and the imaging only shows hyperintense signals on T2-weighted images [10]. However, this is not the case in AD and longitudinal MRIs show a decrease of brain volume instead [9] without any "holes." That means the tissues should restructure as the neuronal deaths increase with time. This leads us to a basic assumption in the proposed model that after the death of neurons, remodeling of the tissue occurs such that the tissue density remains constant while both the mass and volume decrease. We further assume that the atrophy creates internal stress which results in the deformation minimizing the strain energy.

Using Saint Venant–Kirchhoff model for an elastic material, this can be expressed as the minimization of which results in the deformation of the tissue minimizing the strain energy.

$$R(\boldsymbol{u}, p) = \int \mu \mathrm{tr}(\boldsymbol{E}(\boldsymbol{u})^2) + \frac{\lambda}{2}(\mathrm{tr}(\boldsymbol{E}(\boldsymbol{u})))^2 - \int p\,(\nabla \cdot \boldsymbol{u} + a) \tag{15.2}$$

where p is a Lagrange multiplier, μ and λ are Lamé constants, and $\boldsymbol{E}$ is Lagrangian Green strain defined as: $\boldsymbol{E} = \frac{1}{2}\left(\nabla \boldsymbol{u} + \nabla \boldsymbol{u}^T + \nabla \boldsymbol{u}^T \nabla \boldsymbol{u}\right)$.

By taking a sufficiently small time step Δt, this deformation could be reasonably modeled as being linear elastic. For example, for a 2 % global atrophy per year, we have $\Delta t = 1$ year, and the atrophy during the year as $a = 0.02$.

Under linear elastic assumptions, minimizing the energy in Eq. (15.2) is equivalent to solving the following set of equations.

$$\begin{aligned} \mu \Delta \boldsymbol{u} - \nabla p &= (\mu + \lambda)\nabla a \\ \nabla \cdot \boldsymbol{u} &= -a \end{aligned} \tag{15.3}$$

where $\Delta \boldsymbol{u}$ is a component-wise Laplacian of $\boldsymbol{u}$. This equation is very similar to the Stokes flow equation in fluid dynamics. The difference is in the non-zero divergence term which corresponds the loss of mass and volume in the tissue. The momentum equation shows that the gradient of the prescribed volume loss acts as the force term that moves the tissue for the structural remodeling. The Lagrange multiplier p is some sort of virtual pressure whose algebraic values can be seen as the sources and sinks of fluid.

15.2.1 Modeling CSF Region

The timescale of CSF production is hours, which is much smaller compared to the timescale of tissue remodeling due to atrophy. To allow the CSF to expand as required when the brain deforms due to the prescribed atrophy, we release the strict incompressibility constraint in (15.3). Furthermore, the force term of the momentum equation in (15.3) is no longer required. Thus the combined equation for both the brain parenchyma and the CSF regions is:

$$\begin{aligned} \mu \Delta \boldsymbol{u} - \nabla p &= (\mu + \lambda)\nabla f \\ \nabla \cdot \boldsymbol{u} + kp &= -f \end{aligned} \tag{15.4}$$

where we have,

- **Brain parenchyma region:** $k = 0$, and $f = a$
- **CSF region:** $k = 1$, and $f = 0$.

Boundary Conditions: Dirichlet boundary conditions with zero displacement are enforced at the skull.

Material Parameters μ and λ: The deformation model here corresponds to the structural readjustments due to cell loss, thus the Lamé parameters do not have the same usual meaning as during an elastic deformation of the material due to application of an external load/force. The voxel-wise volume change constraint and the boundary conditions, i.e. the shape of the tissue–CSF and brain–skull interface has much more impact on the deformation of the brain parenchyma than any specific scalar values of μ and λ. In the present work these coefficients are set to 1 and 0, respectively.

15.2.2 Staggered Grid Discretization and Finite Difference Method

Equation (15.4) requires a partition of the computational domain into different regions. These regions are obtained by using skull stripping and segmentation of the input baseline brain MRI. The solution of the PDE provides us a deformation field that is applied to the baseline image to generate simulated follow-up image. We use Finite Difference Method (FDM) with staggered grid discretization to solve the system of PDEs in (15.4). Using staggered grid with proper placing of the pressure and velocity variable ensures stability in the solution. FDM is chosen instead of FEM to avoid brain meshing and the complexity of transporting computed variables from mesh to image at each iteration. This allows us to solve the system in a grid that is of the same size as the input image where the grid fits naturally to the image. This also makes it easier to obtain the partition of the computational domain into different regions directly by using a skull stripping and a segmentation algorithm.

For typical brain MRIs of $1\,\text{mm}^3$ resolution, this computational problem size becomes so large that direct solvers are impractical due to memory limitations. The set of equations in (15.4) is similar to Stokes flow equation which is a saddle point system. It needs a suitable combination of an iterative solver and a preconditioner to solve it. We use a Schur factorization to split the equations into the momentum equation and the pressure equation. Each of these equations is solved using different iterative solvers. Our implementation uses composable solvers for multiphysics with PETSc library [1] using `fieldsplit` preconditioner, an approach detailed in [3] with an example for Stokes flow solver with Schur complement factorization. The momentum equation is preconditioned with `hypre` which is an algebraic multigrid preconditioner and can be called from the PETSc interface. The implemented system is run using distributed computing in a locally available cluster.

15.3 Experiments and Results

We use the MIRIAD dataset [13] that has multiple time-point T1 structural MRIs of 45 Alzheimer's patients in the range of 2 weeks to 2 years. Since the dataset contains several time-point scans, we can compare the simulated intermediate time-point images to the corresponding real intermediate images. To prescribe personalized atrophy patterns we need an atrophy estimation for each subject from the extremal time points. We perform the whole brain segmentation using `recon-all` command in FreeSurfer [7]. For the segmentation, FreeSurfer's longitudinal stream [15] is used to create unbiased subject specific templates. This allows us to compare the volumes of large number of regions in the baseline and the follow-up images and estimate atrophy in each of these regions. This estimated atrophy can then be modified and prescribed to each of the baseline MRIs to predict intermediate time-point images. The setup of the experiment we performed is described as follows:

1. Find available extremal time-point scans: baseline I_b and the final follow-up I_f. Let t_f be the time (in years) between the baseline scan and the final scan.
2. Find a mid-point scan I_m that was scanned t_m years after the first scan. This is found by finding t_m that is closest to $t_f/2$.
3. Use FreeSurfer to estimate an atrophy map a_f. This is a scalar image such as the one shown in bottom left of Fig. 15.1. The intensities are the atrophy estimated from FreeSurfer for all the segmented brain regions.

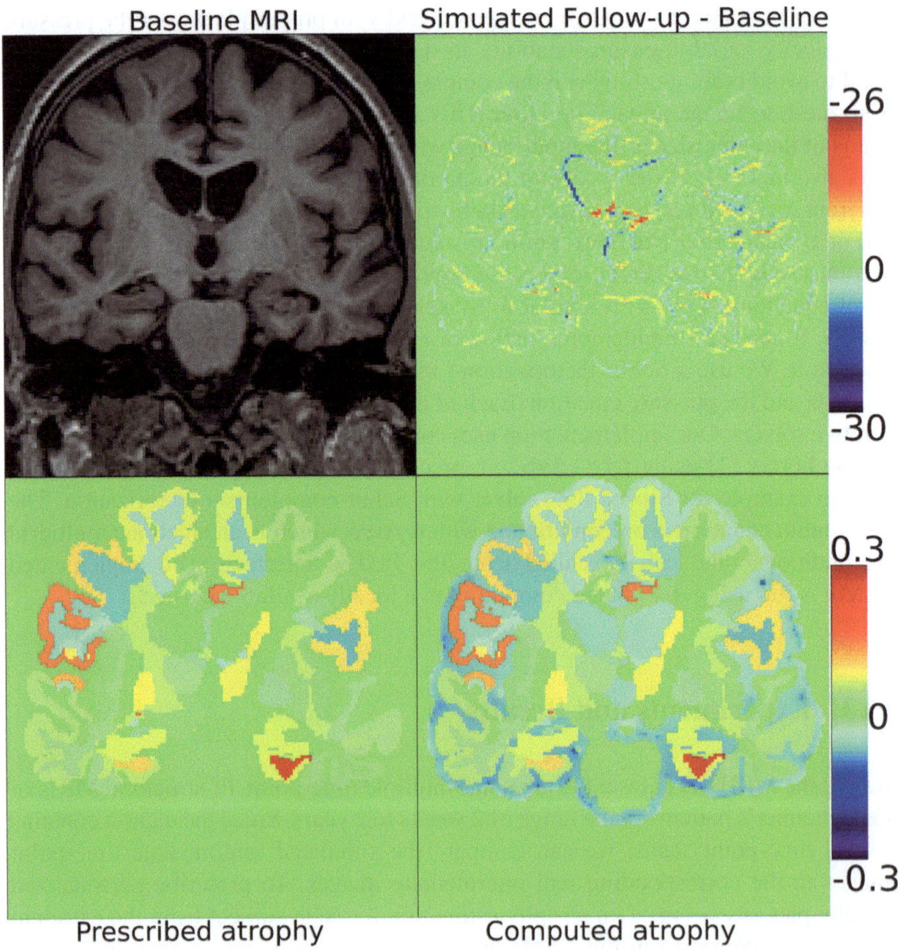

Fig. 15.1 *Top left* is the input baseline MRI to which the atrophy shown in *bottom left* is prescribed. A follow-up image is simulated using our model. The difference of the simulated follow-up and the baseline MRI is shown in *top right*. *Bottom right* is the atrophy map associated to the deformation field that was obtained as the solution of the model when using the atrophy map on the left as input. As expected, in brain tissue region they are the same while in CSF there is expansion to compensate the tissue loss keeping the skull fixed

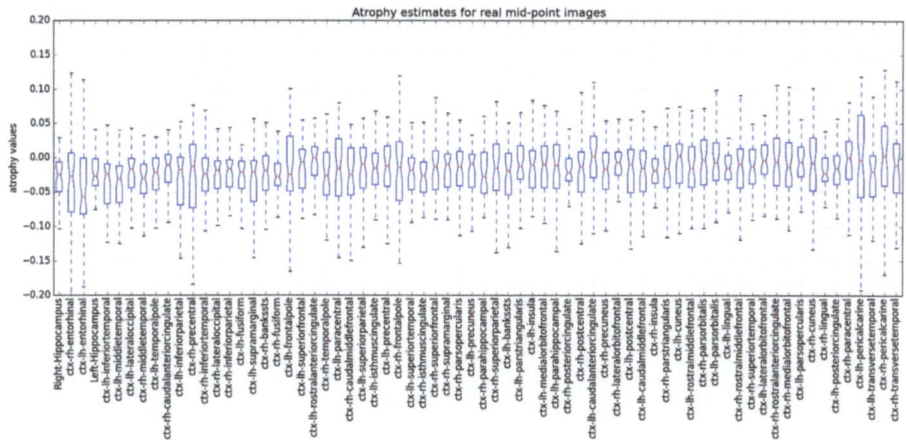

Fig. 15.2 Boxplot of the atrophy estimates for the real mid-point images in the cortical regions and hippocampus. These are the regions that were used in [6]. The data shows the distribution of FreeSurfer atrophy estimates in the AD population of the MIRIAD dataset when considering the first and the mid-point scans

4. Simulate a follow-up image $\hat{I}_m$ that corresponds to the mid-point scan I_m by prescribing a_m where $a_m = a_f * t_m/t_f$.
5. Run the FreeSurfer whole brain segmentation on this simulated image $\hat{I}_m$ and compute volumes of all the segmented regions.
6. Compare FreeSurfer computed volumes of all regions of the images I_m and $\hat{I}_m$.

In Figs. 15.2 and 15.3 we see that for most regions the difference in the atrophy estimation of the interpolated mid-point image and that of actual mid-point image has median close to zero. Higher variability in the difference seems to be mostly in the regions where there is higher variability in the atrophy estimates of the real mid-point images. The large inter-subject variation of the difference between the atrophy estimate in the real mid-point image and the interpolated mid-point image could be due to several reasons. One obvious issue is that the FreeSurfer segmentation with the longitudinal stream expects all the images that are to be segmented to be preprocessed in the same manner. However, in our case the interpolated mid-point image has undergone an extra resampling step while the real mid-point image has not. This extra resampling step is required because the interpolated mid-point image was obtained by warping the real baseline image with a displacement field. Furthermore, the choice of interpolation during the resampling step can also affect the volume measurements by FreeSurfer. We used trilinear interpolation for the resampling. The extra resampling step and the choice of interpolation do have an effect on the estimation of volumes. This has been shown, for instance, in [16] for other segmentation based atrophy estimation techniques. Furthermore, the interpolated mid-point image has a noise (noise inherent in any MRI) that is highly correlated with the real baseline image. However, the noise in

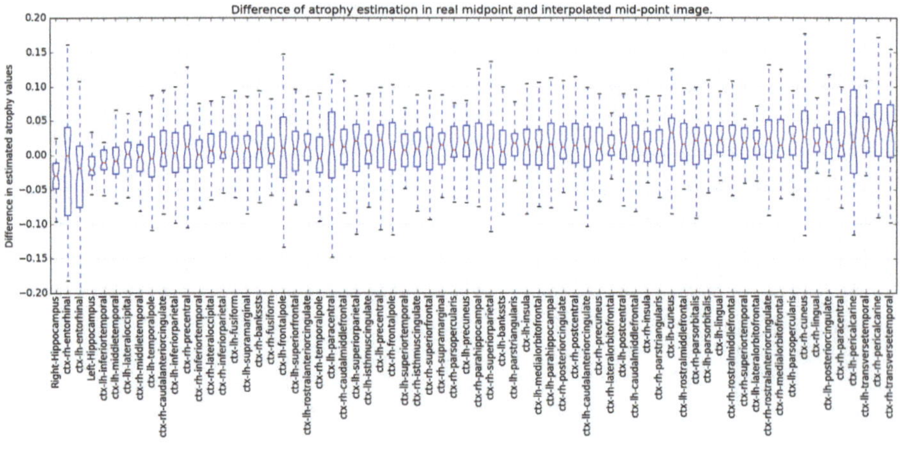

Fig. 15.3 Boxplot of the difference in the FreeSurfer atrophy estimate in the real mid-point image and the interpolated mid-point image for all the 45 AD subjects present in MIRIAD dataset. The regions shown are the same as the one shown in Fig. 15.2 and are displayed in the same order. The interpolated mid-point image is obtained by simulation using the pair of extremal time-point images of each subject

real mid-point image is not correlated to the baseline image. This also affects the atrophy estimation and hence contributes to the variability in the atrophy estimation difference. A detailed analysis must be done to find out the regions that are the most reliable ones in estimating volume changes for both the real and simulated images. The performance of the atrophy measurement tools on simulated images should be thoroughly evaluated to find out the best regions that we can rely upon to test how closely we predict volume changes in new time-point images.

In this case we have interpolated the intermediate time point by linearly scaling the estimated atrophy. For a small time window of a couple of years this is reasonable but if we want to extrapolate, for instance, for several years we would need a non-linear model of atrophy progression. The presented framework allows one to compare the trajectory of brain shape changes with different models of atrophy progression. The ability to prescribe any desired atrophy at any time point allows one to introduce atrophy at different regions of brain at different times. This can be exploited in evaluating the methods proposed in studies such as [8] which order the events from time-series data.

15.4 Conclusions

We have proposed a framework to generate patient specific multiple time-point images based on our biophysical model of brain deformation due to atrophy in AD. The used model is motivated from biomechanical principles and it models the consequence of tissue loss in brain shape changes. From the available two scans of MRI of a patient at two different time points we estimate atrophy in large number of brain structures using FreeSurfer whole brain segmentation [7]. The derived atrophy patterns are linearly scaled and prescribed to the biophysical model to simulate another time-point image. Using the MIRIAD dataset [13] of 45 AD subjects with multiple time points we compare the simulated time-point images against the actual time-point images. The future works will include building the most reliable methods to compare the volumes in simulated and real images. We will also explore the possibility of evaluating methods that study the temporal relationships, ordering, and co-evolution of atrophy in different structures of the brain.

Acknowledgements Part of this work was funded by the European Research Council through the ERC Advanced Grant MedYMA 2011-291080.

References

1. S. Balay, J. Brown, K. Buschelman, W.D. Gropp, D. Kaushik, M.G. Knepley, L.C. McInnes, B.F. Smith, H. Zhang, PETSc Web page (2013), http://www.mcs.anl.gov/petsc
2. H. Braak, E. Braak, Neuropathological stageing of Alzheimer-related changes. Acta Neuropathol. **82**(4), 239–259 (1991)
3. J. Brown, M.G. Knepley, D.A. May, L.C. McInnes, B. Smith, Composable linear solvers for multiphysics, in *2012 11th International Symposium on Parallel and Distributed Computing (ISPDC)* (2012), pp. 55–62
4. O. Camara, M. Schweiger, R.I. Scahill, W.R. Crum, B.I. Sneller, J.A. Schnabel, G.R. Ridgway, D.M. Cash, D.L.G. Hill, N.C. Fox, Phenomenological model of diffuse global and regional atrophy using finite-element methods. IEEE Trans. Med. Imaging **25**(11), 1417–1430 (2006)
5. O. Camara, R.I. Scahill, J.A. Schnabel, W.R. Crum, G.R. Ridgway, D.L.G. Hill, N.C. Fox, Accuracy assessment of global and local atrophy measurement techniques with realistic simulated longitudinal data, in *MICCAI 2007*, ed. by N. Ayache, S. Ourselin, A. Maeder. Lecture Notes in Computer Science, vol. 4792 (Springer, Heidelberg, 2007), pp. 785–792
6. O. Carmichael, D.G. McLaren, D. Tommet, D. Mungas, R.N. Jones, Coevolution of brain structures in amnestic mild cognitive impairment. NeuroImage **66**, 449–456 (2013)
7. B. Fischl, D.H. Salat, E. Busa, M. Albert, M. Dieterich, C. Haselgrove, A. van der Kouwe, R. Killiany, D. Kennedy, S. Klaveness, A. Montillo, N. Makris, B. Rosen, A.M. Dale, Whole brain segmentation: automated labeling of neuroanatomical structures in the human brain. Neuron **33**(3), 341–355 (2002)
8. H.M. Fonteijn, M. Modat, M.J. Clarkson, J. Barnes, M. Lehmann, N.Z. Hobbs, R.I. Scahill, S.J. Tabrizi, S. Ourselin, N.C. Fox, D.C. Alexander, An event-based model for disease progression and its application in familial Alzheimer's disease and Huntington's disease. NeuroImage **60**(3), 1880–1889 (2012)
9. G.B. Frisoni, N.C. Fox, C.R. Jack, P. Scheltens, P.M. Thompson, The clinical use of structural MRI in Alzheimer disease. Nat. Rev. Neurol. **6**(2), 67–77 (2010)
10. R.T. Johnson, C.J. Gibbs Jr., Creutzfeldt–Jakob disease and related transmissible spongiform encephalopathies. N. Engl. J. Med. **339**(27), 1994–2004 (1998)
11. B. Karaçali, C. Davatzikos, Simulation of tissue atrophy using a topology preserving transformation model. IEEE Trans. Med. Imaging **25**(5), 649–652 (2006)

12. B. Khanal, M. Lorenzi, N. Ayache, X. Pennec, A biophysical model of shape changes due to atrophy in the brain with Alzheimer's disease, in *Medical Image Computing and Computer-Assisted Intervention – MICCAI 2014*, ed. by P. Golland, N. Hata, C. Barillot, J. Hornegger, R. Howe. Lecture Notes in Computer Science, vol. 8674 (Springer, Cham, 2014), pp. 41–48
13. I.B. Malone, D. Cash, G.R. Ridgway, D.G. MacManus, S. Ourselin, N.C. Fox, J.M. Schott, MIRIAD–Public release of a multiple time point Alzheimer's MR imaging dataset. NeuroImage **70**, 33–36 (2013)
14. P. Pieperhoff, M. Südmeyer, L. Hömke, K. Zilles, A. Schnitzler, K. Amunts, Detection of structural changes of the human brain in longitudinally acquired MR images by deformation field morphometry: methodological analysis, validation and application. NeuroImage **43**(2), 269–287 (2008)
15. M. Reuter, N.J. Schmansky, H.D. Rosas, B. Fischl, Within-subject template estimation for unbiased longitudinal image analysis. NeuroImage **61**(4), 1402–1418 (2012)
16. S. Sharma, V. Noblet, F. Rousseau, F. Heitz, L. Rumbach, J.P. Armspach, Evaluation of brain atrophy estimation algorithms using simulated ground-truth data. Med. Image Anal. **14**(3), 373–389 (2010)
17. S. Sharma, F. Rousseau, F. Heitz, L. Rumbach, J.P. Armspach, On the estimation and correction of bias in local atrophy estimations using example atrophy simulations. Comput. Med. Imaging Graph. **37**(7–8), 538–551 (2013)
18. A.D.C. Smith, W.R. Crum, D.L. Hill, N.A. Thacker, P.A. Bromiley, Biomechanical simulation of atrophy in MR images, in *Medical Imaging 2003*. International Society for Optics and Photonics (2003), pp. 481–490

Chapter 16
Traumatic Brain Injury: An Investigation into Shear Waves Interference Effects

Grand R. Joldes, Alesio L. Lanzara, Adam Wittek, Barry Doyle, and Karol Miller

16.1 Introduction

Traumatic brain injury (TBI) is one of the leading causes of long-term disability in both industrialised and developing countries around the world. The World Health Organisation states that this type of debilitating injury will exceed many diseases as the major cause of death and disability by the year 2020 [1]. What is this 'silent epidemic' [2] and why is it so damaging? TBI, also known as intracranial injury, results in temporarily or permanently impaired and structural damage to the brain, caused by the application of external mechanical forces to the head. These mechanical forces can be applied through various modes of excitation—rapid accelerations and decelerations, impact loads, inertial loads, blast waves and penetration-by-projectile. The environments these inputs can occur range from the extremely common—road traffic accidents, falls and other unintentional injuries, etc.—to those affecting a particular subset of the population—contact sports, military activity, violence. The associated effects of TBI can range from mild (mTBI), which can lead to cognitive problems such as headaches, memory problems, mood swings and frustrations, to severe, which can lead to major causes of unconsciousness and persistent vegetative state after trauma. The latter is a result of diffuse axonal injury (DAI), involving damage of individual nerve cells (neurons) and splitting of the axonal connection between neurons due to traumatic shearing forces. It should come as no surprise now that there exists a vast amount of literature in the field of TBI, and the research dedicated to reducing its impact ranges from epidemiological studies [1] to the associated costs of TBI [3, 4], to tissue/single

G.R. Joldes (✉) • A.L. Lanzara • A. Wittek • B. Doyle • K. Miller
Intelligent Systems for Medicine Laboratory, The University of Western Australia, Perth, Australia
e-mail: grand.joldes@uwa.edu.au

G. Joldes et al. (eds.), *Computational Biomechanics for Medicine*,
DOI 10.1007/978-3-319-28329-6_16

cell-based work in order to detect the key molecular signatures of the injury [5]. The main problem is the inability to accurately define the relationship between kinetic inputs and subsequent brain injury and its associated effects.

While clinicians and neuroscientists focus on pathological and physiological research, physicists and engineers use the principles of mechanics to study the physical phenomena involved in the TBI process to provide explanations for the cause of brain damage. Various methods have been employed to study the mechanics of brain injury, including animal and human cadaver experiments [6, 7], magnetic resonance imaging (MRI) and elastography (MRE) [8–11], physical modelling [12, 13] and mathematical modelling [14–16]. In particular, finite element (FE) modelling has become paramount in studying the mechanics of brain injury. A critical review of the state-of-the-art brain modelling and simulation for injury prevention together with practical guidelines for analysts creating finite element brain models have been recently published by Yang and King [17].

The stress and strain from FE solutions may be taken as a quantitative measure of tissue damage and correlated with pathological results from clinical and epidemiological investigations [18]. Once good correlations are proven and the FE model is well validated against experimental data specific to the injury mechanism being modelled, it may serve as a valuable tool for better understanding injury mechanisms, injury diagnosis and design of preventive technology.

According to current literature, DAI in humans is estimated to occur at maximum shear strains of 0.1–0.5 and strain rates of approximately 10–50/s [12, 13]. Further studies also suggest that the brain cells are considerably damaged at strains >0.10 and strain rates >10/s [18, 19].

The stresses and strains created by impact loading of the head are the result of dilatational (pressure) and distortional (shear) waves propagating throughout the brain [20]. Dilatational waves exhibit particle motion along the direction of propagation while distortional waves display particle motion transverse to the direction of propagation. The two waves separate over time due to the large difference in wave speeds [21]. It is observed that the dilatational strains are approximately 1000 times smaller than distortional strains, with minor differences between maximum shear strain and maximum principal strains in TBI events [22]. It is also important to note that key membranous structures, the falx cerebri (separating the cerebral hemispheres) and tentorium cerebelli (separating the cerebellum and occipital lobe), are seen to affect the shear wave propagation patterns in the brain due to the change in impedance, encouraging high reflection and attenuation [23].

Upon simulation of an angled frontal impact load to a 3D FE head of MRI resolution, Chen identifies spherically converging shear wavefronts, propagating from the skull boundary towards the inner regions of the brain [24] long after the pressure waves have subsided (Fig. 16.1). What was not investigated nor discussed, however, was the response of the brain in an extended time domain, as travelling shear wavefronts of various frequencies interact with each other after reflection from substructures. These are referred to as interference effects, as superposition of shear wave amplitudes could create localised areas of high shear stress and strain, contributing to the TBI damage mechanism. To date, no direct investigation of shear wave interference has been made in the literature of TBI modelling.

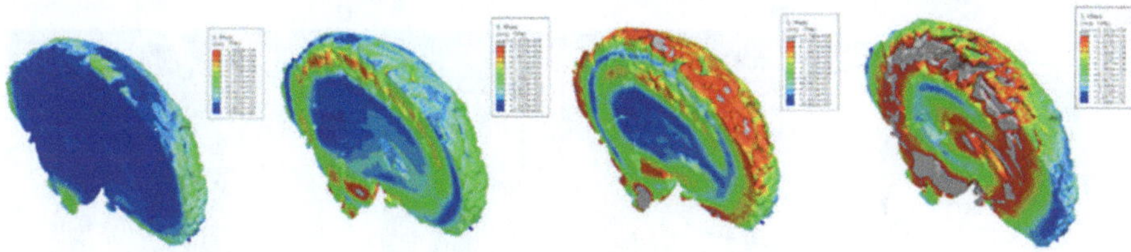

Fig. 16.1 Mid-sagittal view of Von Mises stress distribution depicting spherically converging shear wave propagation (*left* to *right*) over 15 ms

This chapter aims to investigate the effect of shear wave interference as a key mechanism to TBI, by identifying localised regions of the brain exhibiting high strains using a comprehensive FE head model.

16.2 Finite Element Model of the Head

An MRI voxel-based FE mesh of the human head was obtained from Chen [24]. The very fine mesh makes it possible to capture stress wave propagation during impact loading. The model is capable of describing important geometrical features of the head due to the 1.33 mm × 1.33 mm × 1.30 mm voxel size. A Laplacian mesh smoothing algorithm was used to achieve smooth outer surfaces and inner interfaces between tissues. The model's 1,061,799 elements and 1,101,599 nodes are separated into four different tissue types: white matter, grey matter, cerebrospinal fluid (CSF) and skull.

The profiles of the falx cerebri and tentorium cerebelli were carefully sculpted, by reassigning elements from the CSF, grey matter and white matter sets based on the geometry observed in coronal, sagittal and transverse MRI scans. The maximum thickness was two elements (2.66 mm) in the falx cerebri and three elements (3.99 mm) in the tentorium cerebelli, with at least four nodes shared between membrane elements. The thickness used is essential in effectively transmitting bending forces under dynamic simulations and is hence modelled slightly thicker than the approximate membrane thickness of 2 mm [25]. The increased thickness is taken into consideration by scaling the Young's modulus of the membranes in order to obtain the correct bending rigidity (Fig. 16.2).

The material data of the different tissues in the model were taken from [26]. The properties for the introduced falx cerebri and tentorium cerebelli are taken from literature and listed in Table 16.1. All materials are modelled as linear elastic, except the white and grey matter, which are modelled as hyper-viscoelastic (Neo-Hookean with Prony series viscoelasticity).

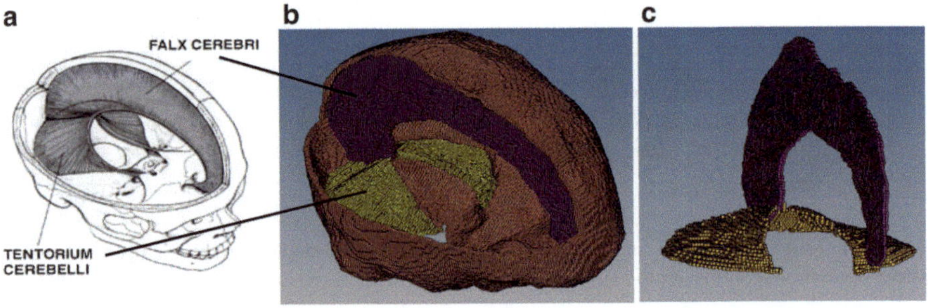

Fig. 16.2 (**a**) The cranial cavity showing the falx cerebri and tentorium cerebelli; (**b**) same view of the structures interior to the skull in the FE model; (**c**) a frontal view of the membranes depicting the tent-like profile of the tentorium cerebelli

Table 16.1 Material properties for falx cerebri and tentorium cerebelli

Tissue	Density ρ [kg/m^3]	Bulk modulus K [Pa]	Short term shear modulus G_0 [Pa]	Reference
Falx cerebri	1130	4.47E+7[a]	4.62E+6[a]	[25]
Tentorium cerebelli	1130	1.32E+7[a]	1.37E+6[a]	

[a]Scaled to account for difference in thickness

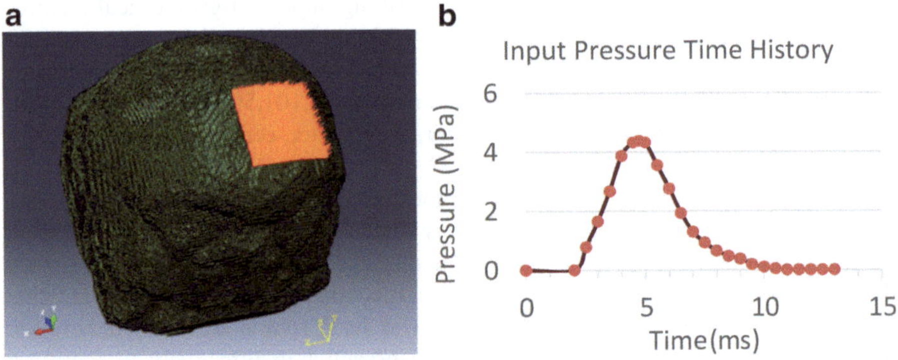

Fig. 16.3 (**a**) The loaded area of the skull. (**b**) Load pressure profile

A fixed boundary condition is used at the head/neck junction. This allows us to capture the rotational motion of the brain, as a free boundary condition would lead mostly to linear motion of the head model. The pulse load shown in Fig. 16.3b is applied to the mid-frontal area of the model in the anterior-posterior direction as a uniformly distributed load over an area of 1556 mm^2, shown in Fig. 16.3a [7].

16.3 Analysis Procedures

16.3.1 Natural Frequency Extraction

The natural frequency extraction step is used to extract the system's modes to be used for the subsequent modal dynamic analysis. The high-performance SIM-based linear dynamics architecture is enabled in this step to ensure that element and material damping factors related to the viscoelastic properties are taken into account in the subsequent modal dynamic analysis. The eigenvalues and the eigenvectors of the biomechanical head system are extracted in this procedure using the Lanczos eigensolver coupled with the SIM architecture [27]. The number of modes used in modal superposition is important in defining accurate dynamic response.

16.3.2 Transient Modal Dynamic Analysis

The transient modal dynamic analysis (TMDA) is used to investigate the shear wave interference process under various input loading profiles and compare the solution to the non-linear explicit dynamic analysis [explicit direct integration method (EDIM)].

While the natural frequency extraction step is computational expensive, finding the solution using TMDA is a relatively inexpensive procedure; therefore it is easy to investigate the behaviour for different loading pressure profiles.

16.3.3 Explicit Dynamic Integration Method

The EDIM is used to investigate the shear wave interference while accounting for non-linear effects. It is ideally suited for large model analyses of high-speed dynamic events such as those seen in TBI. As the equations of motion for the body are integrated using the explicit central-difference integration rule, a large number of small time increments are used [27]. The integrity of the results generated using this method relies on the specified time increment being smaller than the stability limit for the operator, which is based on the highest element frequency in the model and the associated dilatational speeds observed.

16.4 Results

The extracted modes in Table 16.2 show that there is an approximate 10 Hz frequency span between the first 3–4 modes and, thereafter, increasingly smaller increments, until a span of 45 modes from 15 to 60 yields a frequency range of

Table 16.2 Natural frequencies of the head model and the identified modes of interest (in bold)

Mode number	Frequency [Hz]	Effective mass		
		X-component	Y-component	Z-component
1	47.948	6.91E−02	**0.42853**	**1.0844**
2	59.661	3.21E−07	6.87E−05	1.86E−04
3	60.981	**0.176**	4.91E−03	1.59E−02
4	69.707	2.71E−02	9.25E−04	7.81E−04
5	70.873	2.05E−02	**0.12154**	**0.23524**
6	74.858	9.30E−03	8.29E−04	1.25E−04
7	76.616	9.98E−03	1.30E−02	5.13E−03
8	79.156	**0.40435**	6.81E−03	5.64E−03
9	80.436	1.69E−03	1.70E−03	3.82E−02
10	81.132	**0.11532**	5.24E−04	2.16E−03
11	83.051	**0.26127**	5.31E−04	1.34E−02
12	83.685	1.34E−02	1.09E−02	4.75E−04
...	...	...	...	...
60	124.41	4.94E−07	3.62E−03	3.62E−04
Total		2.3304	1.0706	1.7602
Percentage of total mass		8 3%	38 %	63 %

approximately 36 Hz. This is a by-product of the intricately complex structure of the head system. Although there are many complex modes of vibration, each contributing somewhat to the overall response, only a subset of these modes dominates the response of the system under impact loading. This information is contained within the modal effective mass in each kinematic direction, being dependent on the modal participation factors and the modal generalised mass of the system [27].

The effective mass in each kinematic direction for the first 60 modes is highest in the x direction, representing approximately 83 % of the mass of the system. The y and z global kinematic directions return 38 % and 63 % respectively. The number of extracted modes is *not* sufficient to adequately represent the system under a large range of transient inputs or impulse excitation. Approximately 90 % of total mass in each kinematic direction is required for this.

The first natural frequency of the system is 48 Hz, contributing more than one-third of the modal effective mass in the z direction. In fact, of particular interest in the y–z plane are the first and fifth modes for the first 60 modes. Lateral movement is dominated by modes 3, 8, 10, 11, 13 and 15. Figure 16.4 shows some of the associated mode shapes.

It is important to note, however, that due to the rigidity of the skull, stress waves travel much faster here than in the soft tissues of the head. This is well represented by the spherically converging shear waves from impact loads, as a result of indirect loading induced by structural dynamical deformation of the skull. Hence, directionality of impacts may not be as significant in this study.

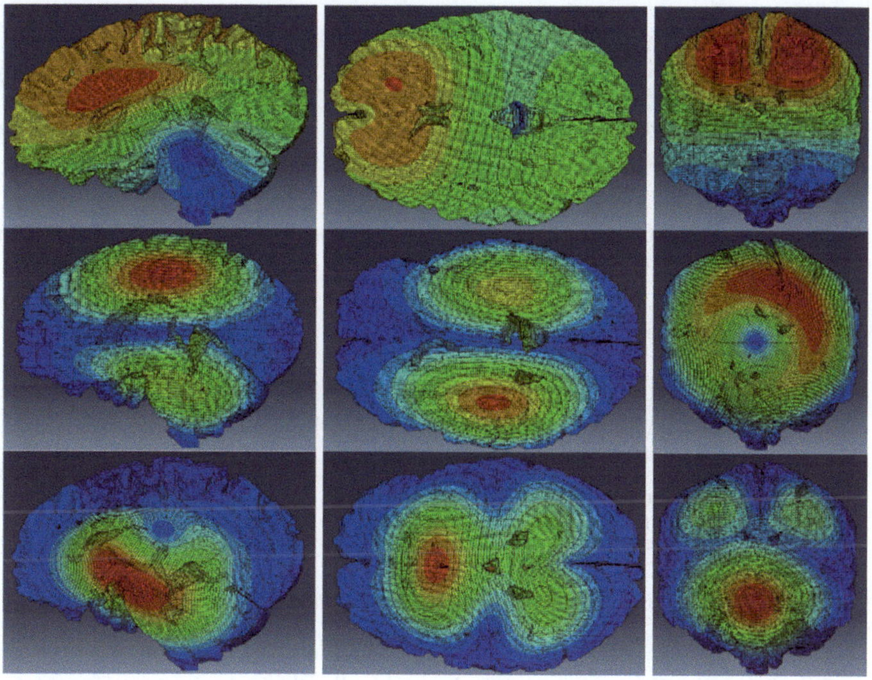

Fig. 16.4 The mode shapes (*left column*) viewed from the mid-sagittal, mid-transverse and mid-coronal cross-sectional cuts of the brain. Regions of largest generalised displacements are in *red*

Table 16.3 Regions of interest identified from the mode shapes

Movement	Mode	Region description	Max. principal strain	% difference from coup
Anterior/posterior	1	Left parietal lobe	0.21	−19 %
	1	Right parietal lobe	0.21	−19 %
		Mid cerebellar region	0.26	0 %
	5	Inferior region of the frontal lobe	0.33	+27 %
Lateral	3	Mid-parietal lobe, directly right of falx cerebri insertion	0.34	+31 %
	8	Cingulate gyrus, close to corpus callosum	0.28	+8 %
	13	Right temporal lobe	0.15	−42 %

Regions of potential interference in brain tissue are identified by analysing perpendicular cross sections (sagittal, coronal and transverse planes) of the 3D generalised displacements and finding areas with the highest displacements. The maximum principal strains for these areas and the comparison with the coup site are presented in Table 16.3.

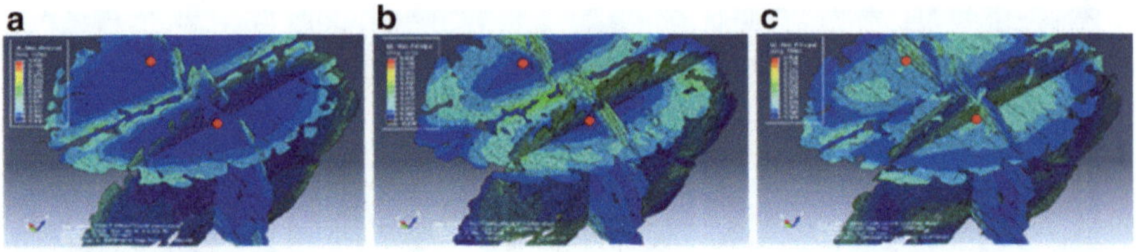

Fig. 16.5 Spherically converging shear waves in both cerebral hemispheres, shown at (**a**) 6.5 ms, (**b**) 8.5 ms and (**c**) 10.5 ms from the EDIM. *Red dots* denote the first two regions of interest

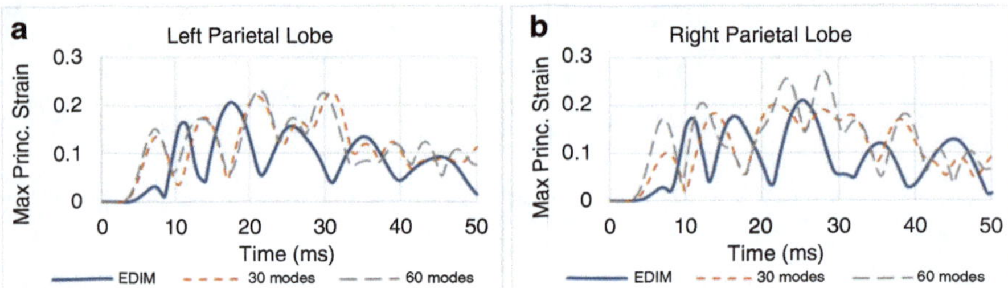

Fig. 16.6 Maximum principal strain responses of (**a**) the *left* and (**b**) the *right* parietal lobes

The dynamic response is mapped graphically through time in Fig. 16.5 to understand wave propagation behaviour. It is observed that wavefronts are not only converging from the outer boundary regions of the brain, but also propagating outwards from the falx cerebri. This creates two converging shear waves in each hemisphere, the focal points of which are the regions identified in the first mode—the mid points of each parietal lobe. The maximum principal strain time history at these two locations is shown in Fig. 16.6. The dominant frequency is ~48 Hz, corresponding to the first natural frequency. Strain amplification is observed, over a longer duration in the right hemisphere. The input load area slightly favours the left hemisphere, which explains the difference in local maximums.

16.5 Conclusions

This study successfully identifies regions in the brain which display higher strains than the coup and counter-coup sites for a frontal impact to the head. The wave patterns are strongly influenced by the falx cerebri and tentorium cerebelli.

The use of modal analysis to identify regions of interference is very effective, by taking into consideration the mode shapes (in strain and displacement) which have strong contributions to the overall response of the system. The TMDA, while not accounting for all solution nonlinearities, does provide an indication of interference effects, as shown by the comparison to the more accurate EDIM solution.

Acknowledgments The authors thank Prof. Martin Ostoja-Starzewski and Ms. Ying Chen from University of Illinois at Urbana-Champaign for providing the mesh of the head.

References

1. A.A. Hyder et al., The impact of traumatic brain injuries: a global perspective. NeuroRehabilitation **22**(5), 341–353 (2007)
2. J.A. Langlois, W. Rutland-Brown, M.M. Wald, The epidemiology and impact of traumatic brain injury: a brief overview. J. Head Trauma Rehabil. **21**(5), 375–378 (2006)
3. I. Humphreys et al., The costs of traumatic brain injury: a literature review. Clinicoecon Outcomes Res. **5**, 281–287 (2013)
4. J.L. Ponsford et al., Costs of care after traumatic brain injury. J. Neurotrauma **30**(17), 1498–1505 (2013)
5. D.M. O'Dell et al., Traumatic brain injury alters the molecular fingerprint of TUNEL-positive cortical neurons In vivo: a single-cell analysis. J. Neurosci. **20**(13), 4821–4828 (2000)
6. D.F. Meaney et al., Biomechanical analysis of experimental diffuse axonal injury. J. Neurotrauma **12**(4), 689–694 (1995)
7. A. Nahum, R. Smith, C. Ward, Intracranial pressure dynamics during head impact, in *Proceedings of the 21st Stapp Car Crash Conference* (Society of Automotive Engineers, Warrendale, 1977), pp. 339–366
8. S. Chatelin et al., Computation of axonal elongation in head trauma finite element simulation. J. Mech. Behav. Biomed. Mater. **4**(8), 1905–1919 (2011)
9. A. Manduca et al., Magnetic resonance elastography: non-invasive mapping of tissue elasticity. Med. Image Anal. **5**(4), 237–254 (2001)
10. M.C. Murphy et al., Analysis of time reduction methods for magnetic resonance elastography of the brain. Magn. Reson. Imaging **28**(10), 1514–1524 (2010)
11. R.J. Okamoto, E.H. Clayton, P.V. Bayly, Viscoelastic properties of soft gels: comparison of magnetic resonance elastography and dynamic shear testing in the shear wave regime. Phys. Med. Biol. **56**(19), 6379–6400 (2011)
12. S.S. Margulies, L.E. Thibault, T.A. Gennarelli, Physical model simulations of brain injury in the primate. J. Biomech. **23**(8), 823–836 (1990)
13. D.F. Meaney, L.E. Thibault, Physical model studies of cortical brain deformation in response to high strain rate inertial loading, in *International Conference on the Biomechanics of Impacts* (Lyon, France, 1990)
14. S. Holm, R. Sinkus, A unifying fractional wave equation for compressional and shear waves. J. Acoust. Soc. Am. **127**(1), 542–559 (2010)
15. R.J. Cloots et al., Biomechanics of traumatic brain injury: influences of the morphologic heterogeneities of the cerebral cortex. Ann. Biomed. Eng. **36**(7), 1203–1215 (2008)
16. P. Chadwick, R.W. Ogden, *A Theorem of Tensor Calculus and Its Application to Isotropic Elasticity* (School of Mathematics and Physics, University of East Anglia, Norwich, 1971)
17. K.H. Yang, A.I. King, Modeling of the brain for injury simulation and prevention, in *Biomechanics of the Brain*, ed. by K. Miller (Springer, New York, 2011), pp. 91–110
18. L. Zhang, K.H. Yang, A.I. King, A proposed injury threshold for mild traumatic brain injury. J. Biomech. Eng. **126**(2), 226–236 (2004)
19. B. Morrison 3rd et al., A tissue level tolerance criterion for living brain developed with an in vitro model of traumatic mechanical loading. Stapp Car Crash J. **47**, 93–105 (2003)
20. A.H.S. Holbourn, Mechanics of head injuries. Lancet **2**, 438–441 (1943)
21. H. Kolsky, *Stress Waves in Solids* (Dover Publications Inc., New York, 1963)
22. D.R.S. Bradshaw, C.L. Morley, *Pressure and Shear Responses in Brain Injury Models*, U.o.S. ISVR, UK, Editor (2001)

23. E.H. Clayton, G.M. Genin, P.V. Bayly, Transmission, attenuation and reflection of shear waves in the human brain. J. R. Soc. Interface **9**(76), 2899–2910 (2012)
24. Y. Chen, M. Ostoja-Starzewski, MRI-based finite element modeling of head trauma: spherically focusing shear waves. Acta Mech. **213**(1–2), 155–167 (2010)
25. N. Yoganandan, *Frontiers in Head and Neck Trauma: Clinical and Biomechanical* (IOS Press, Washington, 1998)
26. Y. Chen, Biomechancial analysis of traumatic brain injury by MRI-based finite element modeling, in *Mechanical Science & Engineering* (University of Illinois at Urbana-Champaign, 2011)
27. DassaultSystèmes, ABAQUS 6.13 Documentation (Providence, 2013)

Chapter 17
Modeling of Bifurcated Tubular Structures for Vessel Segmentation

Haoyin Zhou, Peng Sun, Seongmin Ha, James K. Min, and Guanglei Xiong

17.1 Introduction

Segmentation and geometric modeling of blood vessels from medical imaging is a prerequisite for computational analysis of blood flow and wall mechanics [1], which facilitates the diagnosis of ischemia and atherosclerosis. Currently, vessel segmentation and modeling usually includes two steps: the first step is centerline detection; and with the detected centerline [2], the second step is coronary lumen and wall segmentation [3].

The centerline detection methods mostly start with heuristics-based [4, 5] or learning-based [6] vessel enhancement filtering. Following this, Yang et al. proposed a data-driven centerline tracing method. Zheng et al.'s method is not only data-driven but also assisted with a prior shape model [7], which achieved high accuracy.

Conventional voxel-based coronary segmentation methods are useful for the delineation of vascular geometry. For example, Wang proposed an automatic segmentation method of vasculature that combines level-sets with an implicit 3D model of the vessels [8]. Shahzad et al. performed lumen segmentation by using graph cuts [9]. However, their slow computation speed and inability to incorporate expert knowledge limit their wide use. Learning-based coronary lumen segmentation methods have been proposed for the segmentation of cardiac structures by not only increasing the speed but also learning from the manual annotations. For instance, Lugauer et al. used a learning-based method to segment vessel lumen by using supervised classification [10]. Vessel wall segmentation is much less mature and is primarily performed by global or adaptive thresholding [11]. For the machine learning methods, a water-tight base mesh is required as an initialization for adaption to the unseen image to be segmented. This requirement is difficult

H. Zhou (✉) • P. Sun • S. Ha • J.K. Min • G. Xiong
Weill Cornell Medical College, New York, NY, USA
e-mail: haz2011@med.cornell.edu

G. Joldes et al. (eds.), *Computational Biomechanics for Medicine*,
DOI 10.1007/978-3-319-28329-6_17

for modeling the entire vascular tree, whose topology and connectivity vary from patient to patient. Therefore, to our knowledge, all previous learning-based methods assumed a loose combination of tubular structures and do not account for the bifurcation geometry [6, 7]

In this chapter, we propose a novel method for the construction of complex lumen vasculature with a focus on explicit modeling of bifurcation modeling problem for learning-based vessel segmentation. Compared with those vessel decomposition methods after level-sets [12], there are mainly two steps in our approach. Firstly, the base mesh of bifurcations is modeled by using convex hulls to join extruded tubular structures guided by centerlines. Secondly, subdivision and boosting-based segmentation of the generated convex hull are performed to adapt the model to the target vessel boundaries. Our experiments show that constructed coronary artery models from CT imaging are in high fidelity by comparing to the manual annotated ground-truths.

17.2 Structured Coronary Mesh Generation

Centerlines are widely used to represent the path and connectivity of blood vessels including coronary arteries [13]. In order to characterize the geometry of the lumen boundary, diameters or cross-sectional areas are combined with centerlines for the detection of the presence of any narrowing or stenosis. On the other hand, unstructured meshes (e.g. Fig. 17.1a) are also used to model vessel surfaces, especially for detailed modeling of vascular shape and for generating the fluid domain for computer simulations. Both approaches have limitations considering the tubular and complex shape of blood vessels. Centerlines with diameter information are inadequate to model vessels with asymmetric or noncircular cross-sections.

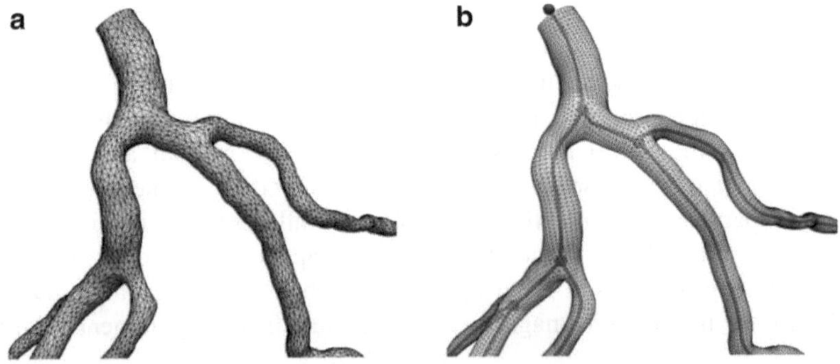

Fig. 17.1 Coronary modeling and segmentation. (**a**) Geometric model of the coronary arteries with an unstructured mesh from the level-sets method. (**b**) The final result of our method: geometric modeling of the same arteries by linking the centerlines with a structured mesh. The centerline endpoints and bifurcation points (shown in *red*) are connected by centerline edges (shown in *blue*)

Although unstructured meshes are excellent for representation of complex surface details, it is challenging to handle queries for global topology and connectivity.

By combining both representations, we propose to model coronary arteries by linking centerlines with structured surface meshes, as shown in Fig. 17.1b. By resampling a given centerline (dense in our case), a list of uniformly distributed nodes is generated as centerline nodes, $c_1, \ldots, c_l$, and a smoother centerline is obtained by finding a spline curve interpolating them. We define a local coordinate frame $[t,u,v]$ at each node c using a rotation minimization technique, where t is along the tangent direction of the centerline and $[u,v]$ spans a 2D plane on the cross-section. The lumen surface is modeled as a structured mesh by sweeping through the contours on the 2D planes. Each contour is implicitly defined using a list of lumen distances, $d_1, \ldots d_k$, from c to the intersections on the lumen surface along k radial vectors sampled uniformly on the circumference.

In our system, we firstly apply the Frangi filter [14] to extract the centerlines. Then, for each center point c, we search for its lumen distances d_k separately by using a boosting-based segmentation classifier. Our boosting classifier is trained from 119 manually annotated patients' data.

17.3 Bifurcation Merging

At bifurcations, as shown in Fig. 17.2a, it is not nature to describe the meshes by centerlines and lumen distances. Getting smooth meshes of the bifurcations, as shown in Fig. 17.2b, can be seen as merging the bifurcating vessels. Our bifurcation mesh generation algorithm includes the following steps: (1) selecting the end-faces of the vessels; (2) generating the convex hull of the end-faces; (3) subdividing and smoothing the convex hull and mesh intersection verifying. The above steps are working in a loop way until the step no intersecting meshes. Finally, (4) boosting-based segmentation is applied to obtain the bifurcation lumen.

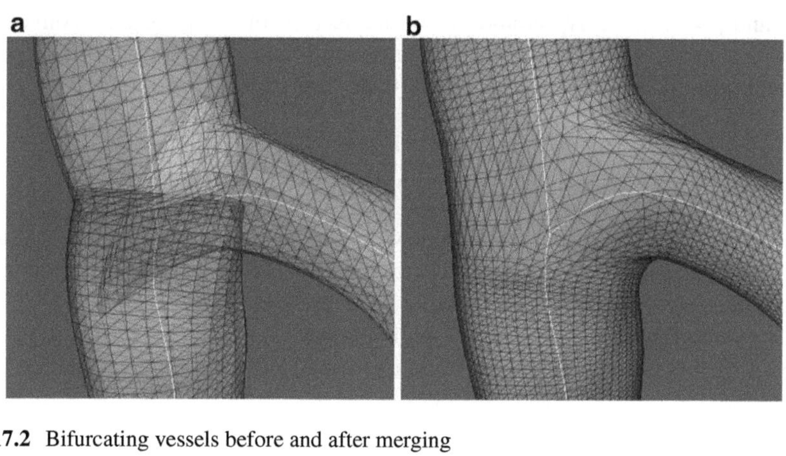

Fig. 17.2 Bifurcating vessels before and after merging

17.3.1 End-Face Selection

The end-faces are the borders between the tubular vessels and the bifurcation. End-faces must be located in appropriate locations relative to the bifurcation point on the centerline because it may be impossible to generate the bifurcation mesh if it is too close or cannot maintain tubular shape if it is further away.

Denote there are M bifurcating vessels and each has N_i contour points, $i = 1, 2, \ldots M$. The contour points are denoted as p_{ij}, where $j = 1, 2, \ldots N_i$.

Condition 1 The selected end-faces of an bifurcation should satisfy that all points $p_{i'j'}$ should be on the same side of end face i, for all $i' \neq i, j' = 1, 2, \ldots N_{i'}$.

Condition 2

$$\left|\overrightarrow{c_i p_{i'j'}} \cdot \overrightarrow{Z_i}\right| > \text{Threshold} \tag{17.1}$$

for $i = 1,..,M$, $i' \neq i$, $j' = 1,2, \ldots N_{i'}$. Where $\overrightarrow{Z}_i$ suggests the unit normal vector of end-face i; c_i suggests the center point of end-face i.

For any end-face that does not satisfy these two conditions, move it far away from the bifurcation center one point at a time on the centerline, until all end-faces satisfy these two conditions.

Condition 1 guarantees that the convex hull exists, and Condition 2 suggests the cases that two mostly parallel end-faces are too close to each other should be avoided.

17.3.2 Convex Hull

A fast and robust algorithm is proposed in this section to obtain the minima convex hull of all M selected end-faces. To generate the mesh, all points p_{ij} for $i = 1, 2, \ldots M$ and $j = 1, 2, \ldots N_i$ should be on the surface of the generated convex hull. Hence, it is necessary to avoid the non-convex end-face cases. Our algorithm obtains the maximum likelihood circle of all end-faces, the radius r_i is obtained by:

$$r_i = \frac{1}{N_i} \sum r_{ij} \tag{17.2}$$

where r_{ij} is the radius of p_{ij}. And the new points generated by r_i are denoted as q_{ij}, $i = 1, 2, \ldots M$ and $j = 1, 2, \ldots N_i$. The convex hull of q_{ij} can be easily transferred into a hull of p_{ij} by replacing the coordinate of the points.

The general idea of our convex hull generation algorithm is that every line segment of this convex hull must have two, and only two adjacent planes. Since

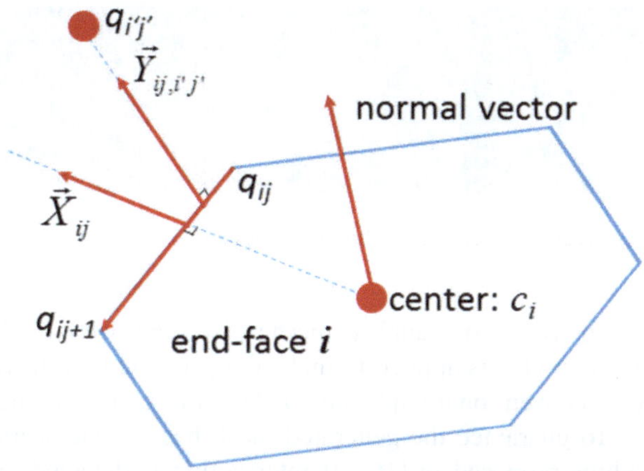

Fig. 17.3 Geometric relationship used in our convex hull generation algorithm

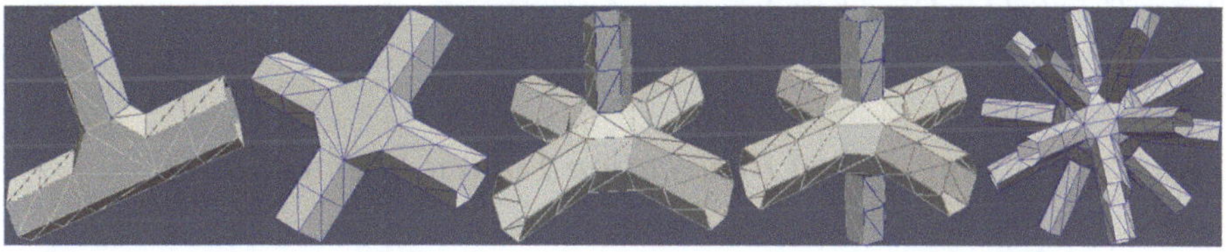

Fig. 17.4 Convex hulls from $M = 3, 4, 5, 6, 20$ bifurcating tubular objects

every line segment $q_{ij}q_{ij+1}$ is already included in the end-face i, one more plane should be found. As shown in Fig. 17.3, for line segment of end-face i defined by points q_{ij} and q_{ij+1}, find $q_{i'j'}$:

$$\text{argmin angle}\left(\vec{X}_{ij}, \vec{Y}_{ij,i'j'}\right), i' \neq i \tag{17.3}$$

where $\vec{X}_{ij} = \overrightarrow{q_{ij}q_{ij+1}} \times \vec{Z_i}$ and points outside, where Z_i is the unit normal vector of end-face i; angle(*) suggests the angle between two vectors

In most cases, the convex plane includes three points: p_{ij}, p_{ij+1}, $p_{i'j'}$. However, sometimes when there are more than four points in this plane, the 2D convex hull algorithms are employed to generate the contour.

After applying this to all line segments, including the new generated line, the final convex hull is obtained. As shown in Fig. 17.4, our algorithm is capable of joining any number of tubular objects by generating the convex hull.

17.3.3 Subdivision and Laplacian Smoothing

To make the generated bifurcation mesh editable, it is necessary to subdivide the convex hull mesh to add some editable points. Here subdivide each line that connects two end-faces into four sub-segments.

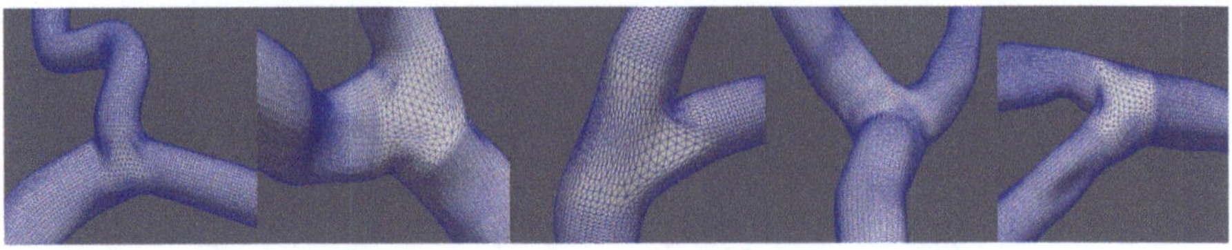

Fig. 17.5 The generated bifurcation meshes

To avoid low quality meshes, an improved Laplacian method proposed by Vollmer [15] is applied to make the generated mesh smooth and nature. Compared with conventional Laplacian method, this can avoid the mesh shrink to the center.

To guarantee the generated mesh has no intersection, we introduced the algorithms proposed in [16]. If intersection is detected, go to Sect. 17.3.1 and use a larger *Threshold* in Condition 2.

17.3.4 Boosting-Based Bifurcation Lumen Segmentation

So far, no CT image information is used in the bifurcation part. To adapt the model to the bifurcation boundaries, we apply the similar classifier as we used in the tubular vessel lumen segmentation. Because the bifurcation mesh does not have centerlines and radius, the boundary is segmented along the normal of the mesh (Fig. 17.5).

The bifurcation lumen segmentation classifier is trained on the manually annotated bifurcation data, here we use 50 patients' data with totally 244 bifurcations and over 1 million points. We use some intuitive features, which is a 25×1 vector includes the CT intensity, intensity · intensity, gradient, gradient · gradient, gradient · normal of itself and its four neighborhood points along the normal.

17.4 Experiment

Coronary CT angiographic images are used to test our algorithms. To evaluate our algorithm, the contracted bifurcations are compared with manually annotated ground truths, as shown in Fig. 17.6. Data from 30 patients are used in our experiment. The data were acquired using standard imaging protocol on GE discovery scanners. Image volumes may contain 153–357 slices, while the size of each slice is the same with 512×512 pixels. For different volumes, the in-slice resolution is isotropic and varies between 0.28 and 0.49 mm with slice thickness from 0.30 to 0.63 mm. The average distance between the constructed bifurcations to the ground truths is $D = 0.299$ mm.

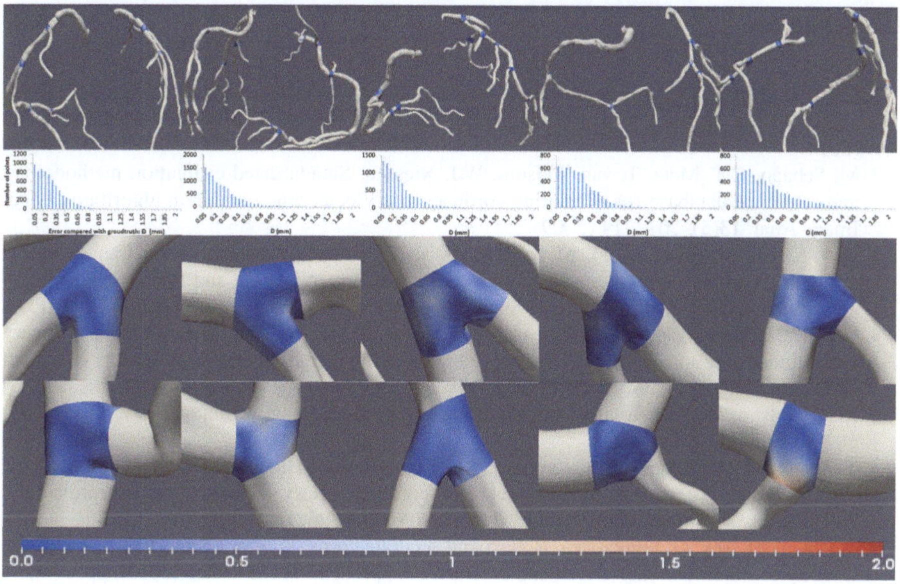

Fig. 17.6 Compared with manually annotated ground truths, the accuracy of our method is evaluated. Patients' coronary vessels are segmented and modeled from CT images as shown in the first row, the bifurcation meshes are colored according to the error distance (D (mm)) between the constructed bifurcations and the ground truths. The distance histograms of D of the bifurcation mesh are shown in the second row. Some examples of detailed colored maps are shown in the third and fourth rows. The last row is the color map parameters of D (mm). Each column represents one patient

17.5 Conclusions

In this chapter, we propose a novel method for construction of complex lumen vasculature with a focus on explicit modeling of bifurcation modeling problem for learning-based vessel segmentation. Convex hulls are used to join tubular structures guided by centerlines. Subdivision and boosting-based segmentation are performed to adapt the model to the target vessel boundaries. Our experiments show that the constructed coronary artery geometry from coronary CT imaging is accurate by comparing to the manually annotated ground-truths. Our future work includes support for interactive editing of the bifurcation geometry and application to other tubular structures, e.g., trachea.

References

1. G.A. Holzapfel, J.J. Mulvihill, E.M. Cunnane, M.T. Walsh, Computational approaches for analyzing the mechanics of atherosclerotic plaques: a review. J. Biomech. **47**(4), 859–869 (2014)
2. M. Schaap, C.T. Metz, T. van Walsum, W.J. Niessen, Standardized evaluation methodology and reference database for evaluating coronary artery centerline extraction algorithms. Med. Image Anal. **13**(5), 701–714 (2009)
3. H.A. Kirisli, M. Schaap, C.T. Metz, T. van Walsum, Standardized evaluation framework for evaluating coronary artery stenosis detection, stenosis quantification and lumen segmentation algorithms in computed tomography angiography. Med. Image Anal. **17**(8), 859–876 (2013)
4. C.T. Metz, M. Schaap, A.C. Weustink, N.R. Mollet, T. van Walsum, W.J. Niessen, Coronary centerline extraction from CT coronary angiography images using a minimum cost path approach. Med. Phys. **36**, 5568–5579 (2009)
5. M.A. Gulsun, H. Tek, Robust vessel tree modeling. Med. Image Comput. Assist. Interv. **11**, 602–611 (2008)
6. Y. Zheng, M. Loziczonek, B. Georgescu, S.K. Zhou, F. Vega-Higuera, D. Comaniciu, Machine learning based vesselness measurement for coronary artery segmentation in cardiac CT volumes, in *SPIE Medical Imaging* (2011), pp. 79621K–79621K
7. Y. Zheng, H. Tek, G. Funka-Lea, Robust and accurate coronary artery centerline extraction in CTA by combining model-driven and data-driven approaches, in *MICCAI* (2013), pp. 74–81
8. C. Wang, R. Moreno, O. Smedby, Vessel segmentation using implicit model-guided level sets, in *MICCAI Workshop* (2012)
9. R. Shahzad, H. Kirisli, C. Metz, H. Tang, M. Schaap, L. van Vliet, T. van Walsum, Automatic segmentation, detection and quantification of coronary artery stenoses on CTA. Int. J. Cardiovasc. Imaging **2918**, 47–59 (2013)
10. F. Lugauer, J. Zhang, Y. Zheng, J. Hornegger, B.M. Kelm, Improving accuracy in coronary lumen segmentation via explicit calcium exclusion, learning-based ray detection and surface optimization, in *SPIE Medical Imaging* (2014), pp. 90343U–90343U
11. D. Dey, V.Y. Cheng, P.J. Slomka, R. Nakazato, A. Ramesh, S. Gurudevan, G. Germano, D.S. Berman, Automated 3-dimensional quantification of noncalcified and calcified coronary plaque from coronary CT angiograph. J. Cardiovasc Comput Tomogr **3**, 72–82 (2009)
12. L. Antiga, D.A. Steinman, Robust and objective decomposition and mapping of bifurcating vessels. IEEE Trans. Med. Imaging **23** (2004)
13. D. Lesage, E.D. Angelini, I. Bloch, G. Funka-Lea, A review of 3D vessel lumen segmentation techniques: models, features and extraction scheme. Med. Image Anal. **13**, 819–845 (2009)
14. A.F. Frangi et al., Multiscale vessel enhancement filtering, in *MICCAI* (1998), pp. 130–137
15. J. Vollmer, R. Mencl, H. Mueller, Improved laplacian smoothing of noisy surface meshes. Comput. Graphics Forum **18**(3) (1999)
16. O. Tropp, A. Tal, I. Shimshoni, A fast triangle to triangle intersection test for collision detection. Comput. Anim. Virtual Worlds **17**(5), 527–535 (2006)

Chapter 18
Modelling the Presence of Diffuse Axonal Injury in Primary Phase Blast-Induced Traumatic Brain Injury

Matthew Sinclair, Adam Wittek, Barry Doyle, Karol Miller, and Grand R. Joldes

18.1 Introduction

Traumatic brain injury (TBI) has become one of the leading causes of death in the modern world [1] and impacts society medically, socioeconomically and emotionally [2]. Blast-induced traumatic brain injury (bTBI) has devastated military personnel since World War I, cloaked under the misclassification of shell shock, among other neurological disorders [3]. Recently, the necessity for understanding the nature of bTBI has become increasingly prevalent due to the insurgence of improvised explosive devices (IEDs) in the Middle East conflicts. The precise mechanism of bTBI continues to evade the full comprehension of researchers due to the difficulty in obtaining accurate in vivo results via human experimentation. The nature and mechanisms that influence bTBI have been extrapolated from several animal-focused tests in combination with interpretation of computer-simulated models and finite element (FE) results. bTBI has become a key focus of military studies as the prominence of IEDs in modern combat continues to threaten the lives of war-fighters around the globe [4].

A primary objective of the investigative community is to determine numerical thresholds that allow for reasonable prediction of the injuries sustained by an individual due to either air or surface blasts. Through analysis of hemispherical surface blast events that are considered survivable on the Bowen survivability curve [5], an attempt at predicting the presence of diffuse axonal injury (DAI) under primary phase conditions can be established and will serve as a foundation for future research in methods of combating this devastating injury. The purpose of this study is to assess the prevalence of DAI in IED scenarios by simulating the effects

M. Sinclair • A. Wittek • B. Doyle • K. Miller • G.R. Joldes (✉)
Intelligent Systems for Medicine Laboratory, The University of Western Australia, Perth, Australia
e-mail: grand.joldes@uwa.edu.au

G. Joldes et al. (eds.), *Computational Biomechanics for Medicine*,
DOI 10.1007/978-3-319-28329-6_18

of primary phase surface blasts using a voxel-based comprehensive computational model of the human head, and comparing the strain values within the brain to tested DAI strain injury criterion. Determination of the presence of DAI in these conditions will serve as a foundation for further study and has potential application within life-saving combat technology research.

Multiple investigations have been performed for air blast scenarios, but only few are accounting for the hemispherical blast surface events. Zhang et al. [6] investigated the influence of blast load dampening via a combat helmet, but only considered small explosives in close proximity to the head model, and only for air blast events. It was discovered that wearing an advanced combat helmet may reduce strain by up to 30 %, but these effects were not accounted for in this study. Explosive impacts on buildings from surface blasts have been studied in [7]. Wang [8] investigated air blast reflection for low mass explosives but primarily focused on bridging vein rupture. To the authors' knowledge, there has been no investigation into the hemispherical surface blast influences on producing DAI using an MRI-resolution head model.

Blast wave trauma has been categorised into four phases: influence from the overpressure waves; collision with shrapnel and flying debris; impact with surroundings from motion induced by blast force; and other factors such as chemical burns or smoke inhalation. These are labelled as primary, secondary, tertiary and quaternary phases respectively [9]. Secondary and tertiary phases are typically compared to standard blunt-force induced TBI, whereas primary phase induced trauma is synonymous with explosive events. The precise mechanisms by which bTBI occurs are still unknown. Grujicic et al. [10] propose that rotational motion and acceleration/deceleration are not applicable in blast-induced trauma scenarios. Conversely, an investigation by Dagro et al. [11] supports the notion that rotational loading is relevant to blast events. In addition, Elder [12] discovered injuries reminiscent of DAI by exposing live animals to blast pressure events, and attributes these injuries to rotational acceleration. Other injury mechanisms are hypothesised, including transmission of the pressure waves via vasculature [13].

Axonal damage, cerebral contusion and subdural haemorrhaging are the three most common forms of mild TBI (mTBI), or concussion as it is known in common language [10]. Of these three forms of mTBI, DAI is the most difficult to detect by conventional means such as computed tomography (CT) scan or magnetic resonance imaging (MRI). DAI occurs when excessive stress or strain is applied to the directional axons within the white matter of the brain, usually resulting from accelerative and declarative forces caused by impact loads. Areas of particular susceptibility include the white/grey matter interface, the brainstem and areas around the falx cerebri. There is no agreement regarding the mechanical and tissue thresholds used for diagnosis of DAI in computational simulations. Experimental data produced by Bain et al. [14] by stretching tissue samples to the point at which predicted morphological injury occurred resulted in DAI threshold strain values ranging from 0.14 to 0.34, with an optimal value of 0.21. This was confirmed via FE analysis by Kleiven et al. [15] whom concluded that a max principal strain of 0.21–0.26 was indicative of DAI through a recreation of both NFL impact events and motorcycle accidents. Although many thresholds have been calculated via computational simulation, in vitro experimentation on human cadavers and human-resembling animal tissue samples remains as a strong basis for DAI threshold detection. For this reason, the empirical principal strain value of 0.21 was selected for use in this study. This threshold was applied to the results obtained from the FE simulation for multiple blast events to determine the presence of DAI.

18.2 Methods

18.2.1 Model and Material Properties

A finite element analysis was performed to determine the strains caused by blast loading within the brain, thereby requiring the use of a computational model of the human head. The model used was a 3D-voxel-based mesh generated from an MRI scanning of an anonymous human brain by Chen [16]. The FE mesh was constructed from hexahedral elements with approximate dimensions of 1.33 mm × 1.33 mm × 1.30 mm. The model consists of white and grey matter, cerebrospinal fluid (CSF) and skull, with material properties derived from Zhang [17]. Custom mesh smoothing algorithms were employed by Chen [16] to provide a better approximate shape of the anatomical features.

The falx cerebri and tentorium cerebellum are extensions of the dura fold separating the hemispheres of the brain and the cerebellum respectively. The original model provided by Chen [16] excluded these anatomical components. Their inclusion was deemed necessary in the simulations in an effort to improve the completeness of the model and to account for their effect on the system response to pressure waves. In addition, several studies empirically determined and reinforced the notions regarding the structural rigidity provided by the falx cerebri and tentorium cerebellum under cranial impacts: Smith et al. [18] demonstrated that the falx cerebri can induce high strains due to impairment of motion of the hemispheres and that the tentorium cerebellum can act as a physical obstruction over which axons can tear, subsequently leading to DAI. In addition, Zhang et al. [19] identify the falx cerebri's strong effect on rotational loading and subsequently, the innocuous effect under translational loading. The falx cerebri and tentorium cerebelli were created by reassigning elements from the CSF, grey matter and white matter sets based on the geometry observed in coronal, sagittal and transverse MRI scans (Fig. 18.1). The maximum thickness was two elements (2.66 mm) in the falx cerebri and three elements (3.99 mm) in the tentorium cerebelli, with at least four nodes shared between membrane elements. The thickness used is essential in effectively transmitting bending forces under dynamic simulations and is hence modelled slightly thicker than the approximate membrane thickness of 2 mm [20]. The increased thickness is taken into consideration by scaling the Young's modulus of the membranes in order to obtain the correct bending rigidity.

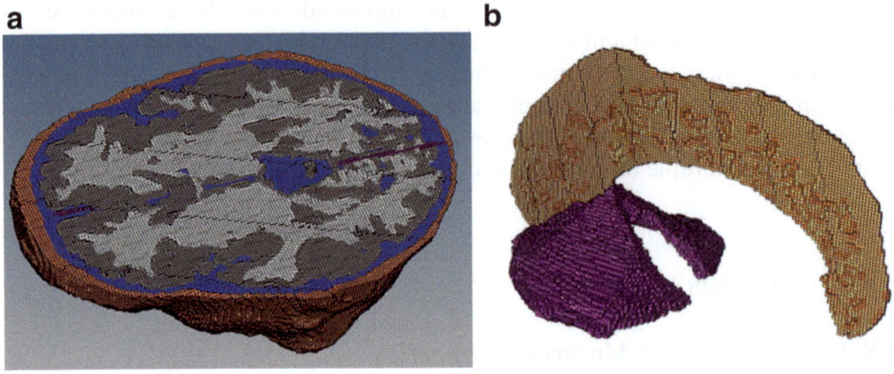

Fig. 18.1 (**a**) Mid-transverse section of the head model. (**b**) The introduced falx cerebri and tentorium cerebellum

Table 18.1 Material properties for falx cerebri and tentorium cerebelli

Tissue	Density ρ [kg/m^3]	Bulk modulus K [Pa]	Short term shear modulus G_0 [Pa]	Reference
Falx cerebri	1130	4.47E + 7[a]	4.62E + 6[a]	[20]
Tentorium cerebelli	1130	1.32E + 7[a]	1.37E + 6[a]	

[a]Scaled to account for difference in thickness

The properties for the introduced Falx Cerebri and Tentorium Cerebelli are taken from literature and listed in Table 18.1. All materials are modelled as linear elastic, except the white and grey matter, which are modelled as hyper-viscoelastic (Neo-Hookean with Prony series viscoelasticity).

A fixed boundary condition was imposed at the stem of the head-spine juncture for this model. This boundary condition allows the introduction of rotational motion of the head, which is a key aspect of producing the strains hypothesised to induce DAI. Chen [16] has analysed the effects of both the fixed and free boundary conditions via comparison with measurements published by Nahum et al. [21]. He determined that the fixed boundary condition provides a reasonable approximation of the rotational acceleration undertaken by the head due to frontal impacts.

18.2.2 IED Simulation

ABAQUS Explicit was used for simulating the explosive events. Two methods of applying the blast loading are available in ABAQUS 6.13: interaction creation with manual amplitude input or use of the Conventional Weapons Effects in Blast Loading (CONWEP) system. The CONWEP system has been used, which calculates a reasonable blast wave inclusive of positive and negative impulses with the input of a TNT equivalence mass and impulse time. Also, the auto-calculation of reflection pressure is especially relevant in this analysis where hemispherical surface blasts are the primary blast loading mechanism.

The guerrilla nature of homemade IEDs makes them crude and variant in their constituents. A "typical" homemade IED is often constructed of 98 % ammonium nitrate and 2 % fuel oil [22]. In combat environments IEDs are usually made of either 105 mm or 155 mm artillery shells which cause more damage to vehicles and convoys, with the 155 mm shell being more explosive than the 105 mm [23]. Their explosive power is characterised by equating their mass specific energy to a TNT mass equivalence [22]. This is achieved through the multiplication of an empirically determined equivalency factor to determine a prediction for expected overpressure and impulse time of a non-TNT based explosion [24]. For ammonium nitrate/fuel oil, 105 mm artillery shell and 155 mm artillery shell, the TNT mass equivalents are 4.5, 2.4 and 7.3 kg respectively. This scaled comparison to an equivalent mass of TNT based on energy output allows for direct comparison between explosives of various constituents.

Using a combination of iterative equations derived by Kingery and Bulmash [25] and the Hopkinson-Cranz distance scaling equation (Eq. 18.1), the appropriate standoff distance Z can be calculated to achieve the given blast parameters including blast overpressure and positive impulse time:

$$Z = \frac{R}{W^{1/3}} \tag{18.1}$$

where R is the actual distance between the detonation and contact points (m) and W is the TNT mass equivalence for the explosive (kg). Table 18.2 details these calculated values and the corresponding blast parameters. Values were chosen to be positioned on the lung damage threshold and 99 % survivability curve [5]. These points were chosen as they demonstrate situations where people exposed to IED blasts would feel the impact, then presumably proceed with their duties without receiving medical attention. It is these scenarios that pose the greatest risk for patients induced with DAI as symptoms of concussions may exist but the full extent of the injury may be underestimated, possibly leading to death. It should be noted that no 2.4 kg explosion was analysed for the 99 % survivability instance. This is because a standoff distance of 1.67 m was required, thereby placing the detonation point above the ground. This reclassifies the explosion as an air blast and hence is not in the scope of this study.

Table 18.2 Simulated blast parameters calculated for IED scenarios

Survivability curve	TNT mass equivalent [kg]	Standoff distance [m]	Reflected overpressure [kPa]
Lung damage	2.4	3.439	506.92
	4.5	4.123	549.20
	7.3	5.329	420.01
99 %	4.5	2.340	2999.13
	7.3	3.413	1557.96

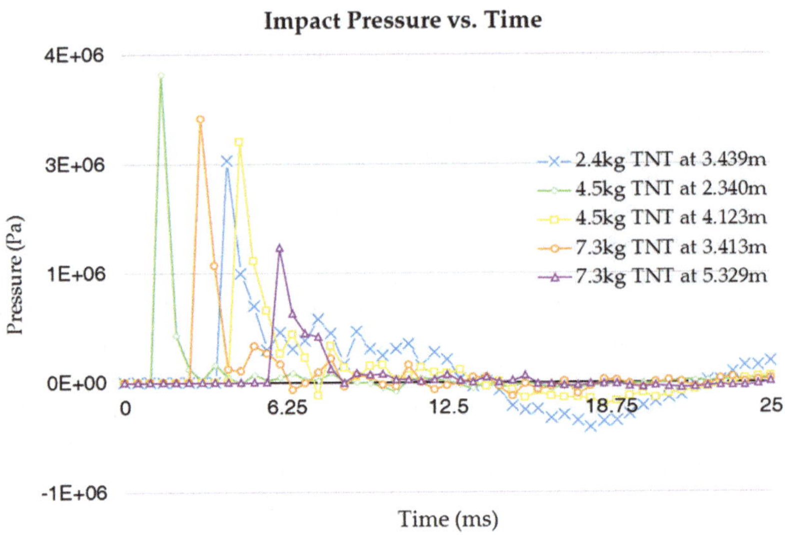

Fig. 18.2 Impact pressure acting on the skull for various TNT mass equivalencies and standoff distances

18.3 Results

The impact pressures, computed at the site of initial impact with the skull, for the various standoff distances and TNT mass equivalences are displayed in Fig. 18.2. Consistent with the survivability curves and expected impacts, the impact pressures depended on the standoff distance and TNT mass equivalence, with the 4.5 kg at 2.340 m having the quickest and highest magnitude impact pressure and the 7.3 kg at 5.329 m conversely having the most delayed and smallest impact pressure. All pressure responses have jagged fluctuations, but the general gradients are similar to those predicted for the ideal Friedlander wave. This is most likely a result of intracranial pressures destructively and constructively interfering with the impact pressure as the results were probed from the impact site on the skull. The sites of impact also varied according to the distance of the blast, suggesting that ABAQUS Explicit has appropriately determined the pressure impact angles, since the closer explosions impacted the head lower on the model. The 2.4 kg explosion at 3.439 m was the only impact pressure that developed a significant negative value, and then increased again above zero. This response was not as expected and could have been caused by a number of potential factors including the particular site of impact or an abnormality in the reflection of pressure waves within the head. Also, for all results, the pressure was analysed directly from the response of the skull, so an oscillating pattern is produced in the impact pressure. This is negligible since the main blast parameter of interest is the peak impact pressure induced.

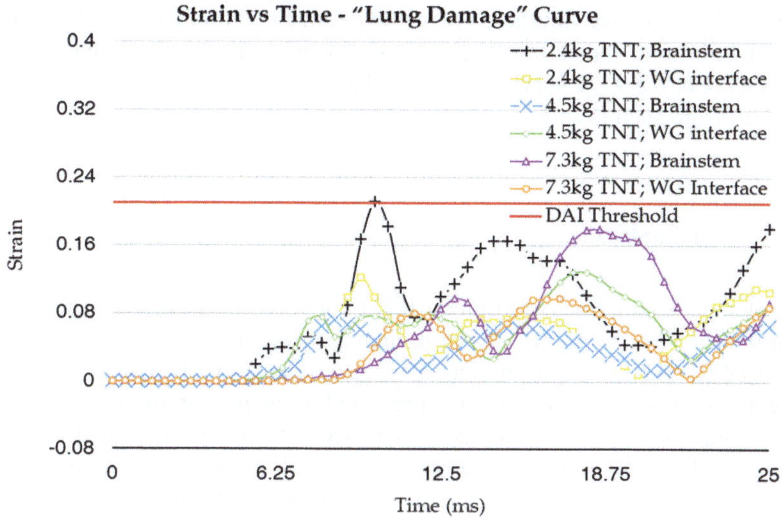

Fig. 18.3 The strain induced by explosions located on the lung damage iso-curve at areas of maximum principal strain in the brainstem and white/grey matter interface

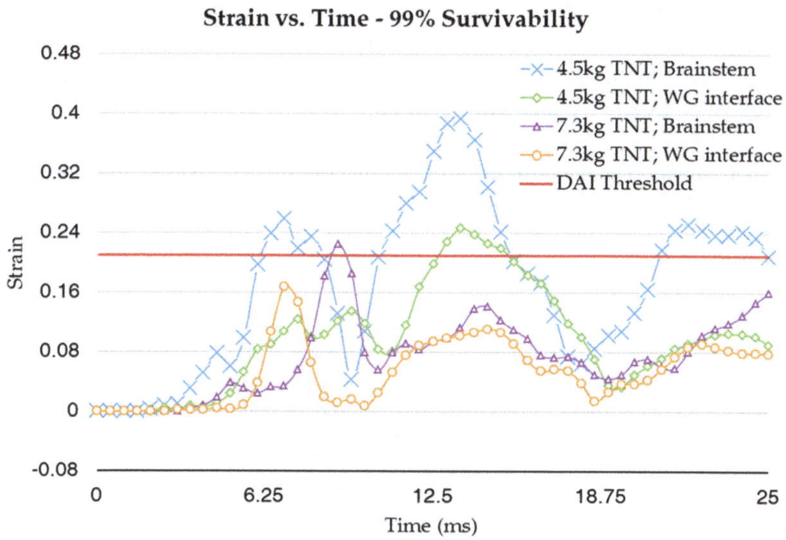

Fig. 18.4 The strain induced by explosions located on the 99 % survivability iso-curve at areas of maximum principal strain in the brainstem and white/grey matter interface

For each of the blast events the response of the model was analysed to determine peak areas of strain (Figs. 18.3 and 18.4). It was noticed in all simulations that high regions of strain were present at the white/grey matter interface and the brainstem. This correlation demonstrates a realistic prediction from the model, as one may see in head trauma events [26].

For the lung damage simulations the maximum strains were generally lower than the DAI threshold, indicating that DAI would generally not be induced by such explosions. For the 99 % survivability simulations there is good indication that DAI will result from such events.

18.4 Conclusions

Finite element simulations of different IEDs, for varying standoff distances according to the 99 % survivability and lung damage iso-curves, were performed using the CONWEP surface hemispherical blast calculations in ABAQUS. The response of the brain was analysed and the predicted principal strains compared to an experimental threshold of 0.21 determined from empirical investigations for the presence of DAI. High regions of strain occurred at locations of the white/grey matter interface and brainstem for all simulations, both areas where DAI will typically be induced.

Strains lower than the threshold were computed for the lung damage curves, suggesting that DAI would not be present in these events. Significantly higher strains were computed for the 99 % survivability events, strongly suggesting that at least mild DAI would be present in these events. This suggests that such an analysis could be used for reasonable prediction of this injury in combat zones without the need for intrusive or expensive medical imaging.

Acknowledgements The authors thank Prof. Martin Ostoja-Starzewski and Ms. Ying Chen from University of Illinois at Urbana-Champaign for providing the mesh of the head.

References

1. A.G. Monea et al., The biomechanical behaviour of the bridging vein-superior sagittal sinus complex with implications for the mechanopathology of acute subdural haematoma. J. Mech. Behav. Biomed. Mater. **32**, 155–165 (2014)
2. J. Topolovec-Vranic et al., Traumatic brain injury among men in an urban homeless shelter: observational study of rates and mechanisms of injury. CMAJ **2**(2), E69–E76 (2014)
3. R.R. Hicks et al., Neurological effects of blast injury. J. Trauma **68**(5), 1257 (2010)
4. N.N. Kleinschmit, A shock tube technique for blast wave simulation and studies of flow structure interactions in shock tube blast experiments (2011)
5. I.G. Bowen, E.R. Fletcher, D.R. Richmond, Estimate of man's tolerance to the direct effects of air blast (1968), report, Lovelace Foundation for Medical Education and Research, Albuquerque NM
6. L. Zhang, R. Makwana, S. Sharma, Brain response to primary blast wave using validated finite element models of human head and advanced combat helmet. Front. Neurol. **4**, 88 (2013)
7. T. Ngo et al., Blast loading and blast effects on structures—an overview. Electron. J. Struct. Eng. **7**, 76–91 (2007)
8. C. Wang, Finite Element Modeling of Blast-Induced Traumatic Brain Injury. Doctoral Dissertation, University of Pittsburgh, 2014

9. G. Ling et al., Explosive blast neurotrauma. J. Neurotrauma **26**(6), 815–825 (2009)
10. M. Grujicic et al., Fluid/structure interaction computational investigation of blast-wave mitigation efficacy of the advanced combat helmet. J. Mater. Eng. Perform. **20**(6), 877–893 (2011)
11. A.M. Dagro et al., A preliminary investigation of traumatically induced axonal injury in a three-dimensional (3-D) finite element model (FEM) of the human head during blast-loading, DTIC Document (2013)
12. G.A. Elder, A. Cristian, Blast-related mild traumatic brain injury: mechanisms of injury and impact on clinical care. Mt. Sinai J. Med. **76**(2), 111–118 (2009)
13. Y. Bhattacharjee, Shell Shock revisited: solving the puzzle of blast trauma. Science **319**(5862), 406–408 (2008)
14. A.C. Bain, D.F. Meaney, Tissue-level thresholds for axonal damage in an experimental model of central nervous system white matter injury. J. Biomech. Eng. **122**(6), 615–622 (2000)
15. S. Kleiven, Predictors for traumatic brain injuries evaluated through accident reconstructions, SAE Technical Paper (2007)
16. Y. Chen, Biomechanical analysis of traumatic brain injury by MRI-based finite element modeling, in *Mechanical Science & Engineering* (University of Illinois at Urbana-Champaign, Illinois, 2011)
17. L. Zhang, K.H. Yang, A.I. King, A proposed injury threshold for mild traumatic brain injury. J. Biomech. Eng. **126**(2), 226–236 (2004)
18. D.H. Smith, D.F. Meaney, Axonal damage in traumatic brain injury. Neuroscientist **6**(6), 483–495 (2000)
19. J. Zhang et al., Role of translational and rotational accelerations on brain strain in lateral head impact. Biomed. Sci. Instrum. **42**, 501–506 (2006)
20. N. Yoganandan, *Frontiers in Head and Neck Trauma: Clinical and Biomechanical* (IOS Press, Amsterdam, 1998)
21. A.M. Nahum, R. Smith, C.C. Ward, Intracranial pressure dynamics during head impact, in *Proceedings of the 21st STAPP Car Crash Conference* (1977), pp. 339–366
22. D. Cormie, G. Mays, P. Smith, *Blast Effects on Buildings* (Thomas Telford, London, 2009)
23. M.B. Panzer et al., Primary blast survival and injury risk assessment for repeated blast exposures. J. Trauma. Acute Care Surg. **72**(2), 454–466 (2012)
24. F. Díaz Alonso et al., Characteristic overpressure–impulse–distance curves for the detonation of explosives, pyrotechnics or unstable substances. J. Loss Prev. Process Ind. **19**(6), 724–728 (2006)
25. C.N. Kingery, G. Bulmash, U.S.A.B.R. Laboratory, Air blast parameters from TNT spherical air burst and hemispherical surface burst. Ballistic Research Laboratories (1984)
26. J. Liu, Z. Kou, Y. Tian, Diffuse axonal injury after traumatic cerebral microbleeds: an evaluation of imaging techniques. Neural Regen. Res. **9**(12), 1222 (2014)

MIX
Papier aus verantwortungsvollen Quellen
Paper from responsible sources
FSC® C105338

If you have any concerns about our products,
you can contact us on
ProductSafety@springernature.com

In case Publisher is established outside the EU,
the EU authorized representative is:
Springer Nature Customer Service Center GmbH
Europaplatz 3, 69115 Heidelberg, Germany

Printed by Libri Plureos GmbH
in Hamburg, Germany